FIELD GUIDE TO SNOW

MATTHEW STURM

Text © 2020 University of Alaska Press

Published by
University of Alaska Press
P.O. Box 756240
Fairbanks, AK 99775-6240

Cover and interior design by 590 Design.
Cover image Stocksy 169598.

All figures and tables by Matthew Sturm unless otherwise noted.

Library of Congress Cataloging-in-Publication Data

Names: Sturm, Matthew, author.
Title: A field guide to snow / [Matthew Sturm].
Description: Fairbanks, AK : University of Alaska Press, [2020] | Includes
    bibliographical references and index.
Identifiers: LCCN 2020003817 (print) | LCCN 2020003818 (ebook)
    | ISBN 9781602234147 (paperback) | ISBN 9781602234154 (ebook)
Subjects: LCSH: Snow.
Classification: LCC GB2603.2 .S78 2020 (print) | LCC GB2603.2 (ebook)
    | DDC 551.57/84—dc23
LC record available at https://lccn.loc.gov/2020003817
LC ebook record available at https://lccn.loc.gov/2020003818

CONTENTS

ACKNOWLEDGMENTS

Over a lifetime of studying snow, I have been mentored, supported and helped by so many people. It was my great good fortune to become Carl Benson's graduate student at the University of Alaska forty years ago and from him I have learned much about snow, about science, and about life. Jon Holmgren and Glen Liston have traveled with me on dozens of winter expeditions, waited patiently in the cold while I fussed about in snow pits, and they have enlightened me on many topics: better companions in science and adventure would be hard to find. Kelly Elder shares my passion for snow pits, stratigraphy, and snow tools; a day spent measuring snow with Kelly is great day, and I have adopted many of his ideas. I have also been blessed with wonderful graduate students over my career, all of whom taught me new things about the snow. Ken Tape, Max Koenig, Simon Filhol, and Charlie Parr deserve special mention, and each of these fine scientists continues to study snow today. Jessica Lundquist and Ken Tape reviewed the book and provided valuable comments that have greatly improved the content. To my many other snow colleagues, ski partners, and friends not named here, I want to say that I have greatly appreciated their insights, good spirits, and friendship while out in the snow. I owe them all a deep debt of thanks. Finally, to my wife, Betsy, and my daughter and son, Skye and Eli, I am forever grateful for their companionship and the amazing patience they have shown while listening to endless mini-lectures on snow.

PREFACE

I have spent forty years of my life in the scientific study of snow, examining how it impacts our climate, water supply, the animals that paw through it or burrow under it, and the trees, shrubs, and tundra that it covers as it falls. Over the same period, living in Alaska, I have spent thousands of hours skiing, snowshoeing, snowmobiling, and shoveling my driveway. Some might think mine has been a 'cold' life, but it has been wonderful.

That life has given me ample time to observe a wide variety of human reactions to snow, reactions that range from fear and hate, to joy and love. We curse snow on the road if it makes us late to an appointment, but we love a "snow day." A recent poll

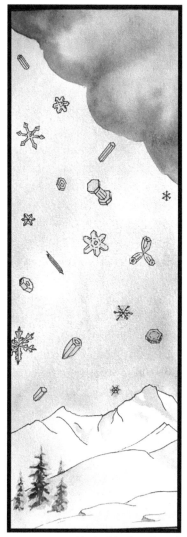

(Art by Matthew Sturm)

we conducted found 73% of respondents had a positive view of snow, which is heartening to a snow scientist. When asked, people recounted with fondness things they had done in the snow like building a fort, sledding, or walking through snowy woods. A common response was to recount with relish a personal snow epic, for instance, being trapped by a blizzard in a cozy cabin. Virtually everyone thought snow was beautiful. But the same survey revealed people had little knowledge of how snow worked, the science of snow. Knowing that aspect of snow can enrich people's appreciation of snow even more.

Over my career I have led more than thirty winter expeditions, taught snow science in the classroom at both undergraduate and graduate levels, and run more than a dozen field method classes. In the field with my students, we have worked shoulder-to-shoulder in snow pits measuring the density, grain size, and stratigraphy of the snow pack. In that setting I have tried to teach how to read in the snow the story of the winter. Beginners, at first, have trouble seeing that story: snow is all white and the same. But with time, and a little guidance, the winter world begins to open for them. That is what this book is about: a little guidance for those who might not get a chance to dig a snow pit with me, but who want to know more about the snow. I hope readers will find it useful and informative.

The field guide:

1. Begins with an overview of where snow is found on the Earth.

2. Then looks at how clouds produce snowflakes.

3. Next, the guide explores how these wonderful ice gems get down to the surface of the Earth and why they take the shapes we see.

4. The guide then examines how a combination of periodic snowfall events plus weather (wind, sun, temperature) combine to create the diverse layers of snow that make up the snow pack.

5. Next, the guide explores how these layers, once deposited, begin to change rapidly (metamorphism) due to the physics of ice.

6. The guide then examines how these snow layers and specifically their optical, physical, and mechanical properties, impact our planet, our climate, the plants and animals of our world, and us.

7. The guide ends with a discussion of snow and living things: animals, plants, and people.

Looking up from below at depth hoar cups at the base of the snow. (Photo by Matthew Sturm)

1

WHERE IS SNOW?

Other than air, water, soil and rocks, and perhaps clouds, it is hard to think of any other substance as widespread as snow. In a good winter, snow covers 50 million square kilometers (19 million sq. miles) of the Northern Hemisphere, nearly 50% of the total land area (Fig. 1-1). Snow covers the ice of the Arctic Ocean and the freshwater ice on many of the large northern lakes that freeze in winter (e.g., Lake Winnipeg), adding 17 million square kilometers. So all told, more than 67 million square kilometers (27 million sq. miles) of the Northern Hemisphere get covered by snow each winter. In some locations that snow lingers for nine months of the year, making snow the predominant surface cover of the Earth in those places. The Southern Hemisphere, due to its geography, hosts considerably less snow, perhaps another few million square kilometers, but nearly all of the Antarctic continent is snow covered, adding an additional 14 million square kilometers (5.4 million sq. miles) there to the overall total. These numbers suggest that ours is a very snowy world.

Even in warmer places where snow is unlikely to stick and pile up, it can still be seen falling from the sky (Fig. 1-2). Over vast areas of the Earth, more than a third of the total annual precipitation falls as snow. Snow has even occasionally been seen falling in Florida, where in Panama City on December 22, 1989, it snowed a full two inches.

Week 51 Snow - 2009

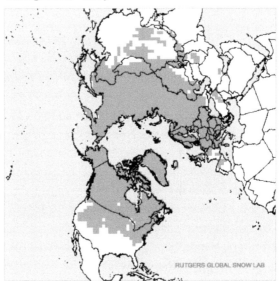

FIGURE 1-1. The winter of 2009–2010 was a good year for snow. It covered over 50 million square kilometers of the Northern Hemisphere (tan area), extending south nearly to the Mexican border. In most years there is a bit less coverage, but the amount always exceeds 40 million square kilometers. Unlike many other aspects of the cryosphere (e.g., permafrost and sea ice), which are disappearing rapidly due to global warming, the annual extent of seasonal snow appears to be increasing slightly. NASA has been tracking the global coverage of snow since the 1970s using satellites, and the data can be viewed at https://climate.rutgers.edu/snowcover/chart_vis.php?ui_year=2009&ui_week=51&ui_set=0))1-1.
(David Robinson, Rutgers Global Snow Lab. Used with permission.)

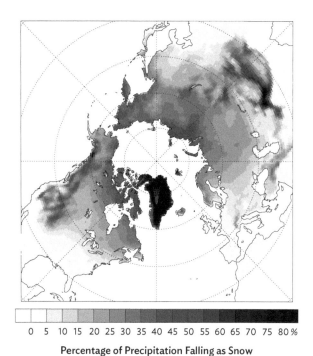

0 5 10 15 20 25 30 35 40 45 50 55 60 65 70 75 80 %

Percentage of Precipitation Falling as Snow

FIGURE 1-2. The percentage of annual precipitation (rain plus snow) that falls as snow based on re-analysis of global climate data. (Analysis courtesy of D. Slater.)

In Europe, falling snow is common as far south as Spain and can be found at the equator, where the tall peaks of New Guinea, Ecuador, and Tanzania not only receive snow, but enough snow to produce glaciers.[1]

So opportunities for observing snow, either on the ground (where it is called the *snow cover* or *snowpack*), or as falling crystals and snowflakes, are quite widespread. In fact, there are very few places on Earth where an interested observer cannot engage with snow in some way.

It is no surprise then, that due to accessibility, their astounding symmetry, and their great beauty, that snowflakes have been observed and photographed for both science and fun for over seven hundred years (Fig. 1-3 and Table 1-1). More on snowflakes and crystals appears in Chapter 2 of this guide, but excellent resources can be found online and in print that can help a beginner learn how to observe, classify, and photograph snow crystals.[2]

Snow on the ground also has something of a cult following, in large measure because of the fun of snow recreation rather than the "fearful symmetry" of snowflakes. In 2013, nearly 150,000 snowmobiles were sold worldwide, and it is estimated that there are 330 million downhill skiers out there fueling a $14 billion/year industry. Some of these fanatics spend a considerable amount of their time parsing weather reports and ski area websites to learn how much base snow is present and whether appreciable new snow (powder) has fallen. Ski areas where low-density new snow falls frequently become famous for their powder skiing, and

---

[1] https://principia-scientific.org/equatorial-glaciers-yeah/
[2] Learn more about snowflakes and crystals online at snowcrystals.com and https://www.cbc.ca/natureofthings/features/dons-guide-to-taking-photos-of-snowflakes.

Snow Crystals

FIGURE 1-3. Snowflakes observed and sketched by Captain William Scoresby while whaling near Greenland, 1810–1820. The figure comes from Scoresby's memoir (1820). Prior to this practical whaler's careful sketches, many fanciful snow crystals had been drawn and appeared in books, including ones with eight rather than six arms.

William Scoresby from *An Account of the Arctic Regions with a History and Description of the Northern Whale-Fishery: The Whale-Fishery.*

TABLE 1-1. Snowflake studies through the ages

Year	Author and Accomplishment
1250	Albertus Magnus: oldest known detailed description of snowflakes.
1555	Olaus Magnus: earliest diagrams of snowflakes in *Historia de gentibus septentrionalibus.*
1611	J. Kepler, *Strenaseu De Nive Sexangula,* explains hexagonal symmetry.
1637	R. Descartes's *Discourse on the Method* includes hexagonal snowflake diagrams.
1660	E. Bartholinus, *De figura nivis dissertatio,* sketches snow crystals.
1665	R. Hooke observes snow crystals under magnification in *Micrographia.*
1675	F. Martens, a German physician, catalogues twenty-four types of snow crystals.
1681	D. Rossetti categorizes snow crystals in *La figura della neve.*
1778	J. Florentius Martinet diagrams precise sketches of snow crystals.
1796	Shiba Kōkan publishes sketches of ice crystals under a microscope.
1820	William Scoresby's *An Account of the Arctic Regions* includes snow crystals by type.
1832	Doi Toshitsura describes and diagrams eighty-six types of snowflakes.
1837	Japanese scientist Suzuki Bokushi publishes *Hokuetsu Seppu* showing snowflakes.
1840	Doi Toshitsura expands his categories to include ninety-seven snowflake types.
1855	J. Glaisher publishes detailed sketches of snow crystals under a microscope.
1865	F. E. Chickering publishes *Cloud Crystals – a Snow-Flake Album.*
1872	J. Tyndall publishes *The Forms of Water in Clouds and Rivers, Ice and Glaciers.*
1891	F. Umlauft publishes *Das Luftmeer.*
1893	G. Hellmann and R. Neuhauss photograph snowflakes under a microscope in *Schneekrystalle.*
1894	A. A. Sigson photographs snowflakes under a microscope.
1902	W. Bentley publishes his famous snowflake images in *Weather Bulletin* (200 photo-micrographs).

locals develop nicknames for their best snow (e.g., Bridger Bowl's "cold smoke" in Montana or Utah's "powder paradise." (For more on this mania, see *Secrets of the Greatest Snow on Earth* by Jim Steenburgh [2014]).

But snow on the ground is also so common that people tend to take it for granted. Outside of a few snow scientists making research observations, those measuring the snowpack generally do so for practical considerations, such as ensuring adequate sources of water and hydropower. Billions of people worldwide get their drinking and agricultural water from snowmelt, yet only a fraction of these people are aware of that fact, and few ever go out and measure the snow on which they rely. One group that is an exception to this rule are backcountry skiers and snowboarders, who dig snow pits and make snow observations in order to assess the snow stability and minimize their risk of being caught in an avalanche, as well as to know whether it is worth skinning up a mountain in order to ski down.

Of course, falling snow and snow on the ground are intrinsically linked, the first producing the second through a series of snowfalls and subsequent weather events, a connection explained in detail in Chapter 3.

Since both the amount of snow and the nature of the weather that produces the snow vary in distinct ways with geography, we can subdivide snowpacks into six distinct climate classes (Fig. 1-4), using a system devised in 1995. The snow class names come from vegetation biomes (tundra, boreal forest, montane forest, prairie) and from Andre Roche's 1949 avalanche regime classification system (which had alpine and maritime classes), but the classes themselves are actually defined by the number and character of the snow

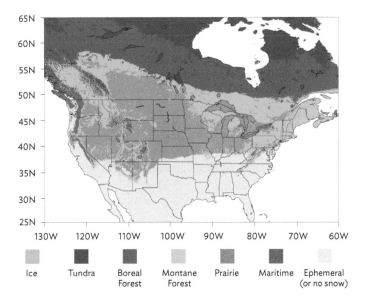

Ice   Tundra   Boreal Forest   Montane Forest   Prairie   Maritime   Ephemeral (or no snow)

FIGURE 1-4. The general distribution of snow climate classes across North America. The types of snow layers that define each class are shown in Figure 1-5. The geographic distribution of the classes changes through the winter season and with the local weather, so it is perfectly possible, for example, to have an ephemeral snow cover in Denver in early winter, then build up a prairie snow cover, which eventually transitions to a montane forest snow cover when enough additional snow has fallen.

(M. Sturm, J. Holmgren, and G. E. Liston, "A Seasonal Snow Cover Classification System for Local to Global Applications," *Journal of Climate* 8, no. 5 (1995): 1261–1283.)

layers that make up each snowpack. Because the classes are defined this way, the snow class is not really about where the snow is located, but rather what it is comprised of, an aspect of the classification system that many people (including some of my fellow snow scientists) have misunderstood. In fact, it is perfectly possible for a tundra snowpack to form in an open, windy field near Lake Tahoe, an alpine snow cover

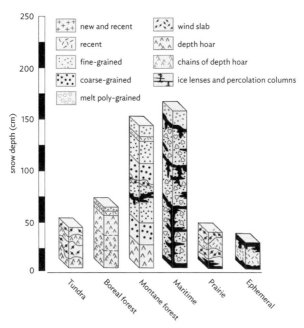

FIGURE 1-5. The snow classes are defined by the number, character, and in some cases, the thickness of the layers that make up the snowpack. For example, maritime snow tends to be deep and always has a lot of icy features in it because rain and/or melt water have percolated down into the snowpack and refrozen. Tundra snow is always shallow and has few layers, invariably including some wind slab layers.

(M. Sturm, J. Holmgren, and G. E. Liston, "A Seasonal Snow Cover Classification System for Local to Global Applications," *Journal of Climate* 8, no. 5 (1995): 1261–1283.)

to develop in upstate New York, or a prairie snow to develop in a windy Ohio cornfield. Figure 1-5 shows the "typical" snow layers found in each class. The symbols depicting these layers are explained in more detail in Chapter 5, where we take an in-depth look at snow layers.

In the following chapters, metamorphic processes and principles related to snow are introduced that are applicable across all of these snow classes. We will see that for a given class, there will be some processes that are more dominant than others. But a close look at the map will make it clear that a motivated observer in the U.S. can get to several different classes of snow if they want to without excessive travel—and because the class of snow changes with time as the winter season waxes and wanes (Fig. 1-6), simply being patient, an observer will be rewarded with the opportunity to see many diverse snow features without traveling far.

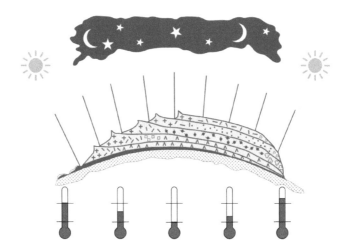

FIGURE 1-6. The snowpack evolves throughout the winter year, the state of the snow being governed by temperature, sunlight and darkness, wind, and snowfall.

2

THE CRYSTAL FACTORY

When I was a kid, I would try to catch falling snowflakes on my tongue. Mouth open, I ran around looking up at the clouds from which the snowflakes were falling, with absolutely no idea what complex physics were taking place up there in order for it to snow. Now, when I think of those clouds, I think of them as billowy factories producing millions of tiny ice masterpieces (Fig. 2-1).

A cloud is made of water vapor, water droplets, and if it is cold enough, small ice crystals. The crystals and droplets can condense directly from the water vapor (a gas), but the rate at which they can do so is so slow that it is certainly

FIGURE 2-1. Exquisite tiny ice masterpieces produced in a cloud.
(Image courtesy of Ken Libbrecht, one of today's masters of snowflake photography.)

not the main mechanism that produces the snow and rainfall on Earth. A much faster mechanism must be at work. That mechanism is one in which the water vapor condenses initially on tiny foreign particles (called cloud condensation nuclei, or CCNs). These can be particles of dust, sea salt, soot, sulfate, or even pollen and bacteria. CCNs lower the threshold conditions for vapor condensation, acting as catalysts, and by that means, greatly speeding up the rate of formation of raindrops or snowflakes.

Marine sulfate and sea salt CCNs get up into the clouds when the wind whips up ocean waves (Fig. 2-2a) or when bubbles burst at the sea surface and spray particles into the air. Soot particle CCNs come from forest fires and smoke stacks (Fig. 2-2b). Other CCNs come from volcanic eruptions. Recently, scientists have even found minute organic plant material floating about in the atmosphere, material that has proven particularly effective at nucleating snowflakes. The CCNs are small, about 0.1 micron (μm) across (1/10,000th of a millimeter) (0.004 thousandths of an inch).

Some readers may have heard of cloud seeding, a process to make it rain. This process was developed by Dr. Irving Langmuir and his group at General Electric in the late 1940s (see *The Brothers Vonnegut* by Ginger Strand [2015]). Cloud seeding is accomplished by introducing CCNs into a cloud using finely powdered silver iodide that is burned and the smoke sprayed from an airplane or lofted from a ground-based tower sprayer. The silver iodide, like salts and sulfate, is a good nucleator of water droplets, and it increases the likelihood of nucleating snow particles. The practice itself remains controversial, however, because its efficacy (thought to be about a 10% increase in precipitation) is difficult to establish.

FIGURE 2-2a. An ocean wave lofts salt spray and sulfate into the air, both sources of CCNs (cloud condensation nuclei) on which cloud droplets nucleate.
(https://www.needpix.com/photo/download/1311431/wave-ocean-nature-beach-sea-curl-surf-shore-water.)

FIGURE 2-2b. A forest fire billows soot into the atmosphere, which along with dust and bacteria, make good condensation nuclei.
(https://www.alaskapublic.org/2018/08/14/whats-going-on-with-alaskas-17-wildfires/.)

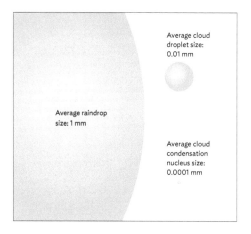

FIGURE 2-3. The relative sizes of raindrops, cloud droplets, and condensation nuclei. Table 2-1 gives the volume for these same drops assuming they are spherical. A raindrop is 6 orders of magnitude (a million) times larger in volume than the initial cloud droplet from which it is derived, so it takes the coalescence of a whole lot of droplets or ice particles to make just one raindrop or snowflake.

But even with CCNs present, there remains a physics problem in producing snow or rain: the vapor pressure of an ice particle or a water droplet (a measure of how quickly the liquid will evaporate) is a very strong function of its size (or how tightly its surface curves), something called the **Kelvin effect**. The smallest droplets (Fig. 2-3) have the highest vapor pressures, and because of this, in most conditions in a cloud they tend to disappear before they have time to grow. So how then do these tiny droplets or ice crystals get past that hurdle and get big enough to avoid evaporating away?

A second physics effect, described by **Raoult's Law**, comes into play. This effect describes how a droplet's vapor pressure is reduced if any dirt or impurities are present in the droplet (a solute effect). Since in a cloud there are always

TABLE 2-1. CCN and droplet sizes and volumes

	Diameter (mm)	Volume (mm³)	Number (per cm³)
CCN	0.0001	5.24×10^{-13}	1,000
Haze droplet	0.001	5.24×10^{-10}	1,000
Cloud droplet	0.01	5.24×10^{-7}	1,000
Drizzle droplet	0.1	5.24×10^{-4}	1
Rain droplet	1	5.24×10^{-1}	0.001

impurities, then to some degree Raoult's Law is always in operation. The rule competes against the Kelvin effect, with the combined result of the two being a dynamic balance where a droplet or ice crystal might grow or evaporate depending on its initial size and how saturated the cloud is. The CCNs help here to get the process started by initiating the condensation, but in the end it is actually the movement of air masses that matters most, the updrafts and convective air movement in the cloud that make the whole thing work.

By simply forcing a cloud parcel upward (an updraft), we can drop its temperature without the loss of any water vapor. When we do this, the air cools off, and a cloud goes from being saturated to **supersaturated**. That does the trick for getting those tiny particles to grow rather than fade away. With the cloud supersaturated, the particles are on their way to growing into raindrops or snowflakes. A night fog forming above a lake is a graphic illustration of this updraft-cooling-supersaturation linkage. The air above the water is saturated, and as night falls and the temperature drops (which is equivalent to lifting an air parcel higher and cooling it), the air becomes supersaturated, and then condensation occurs. A fog appears hovering over the water as if by magic.

There is a second, even more efficient way, to make snow crystals rapidly. It requires understanding that water can be supercooled. That means we can bring the temperature of liquid water down below freezing and it will still be a liquid. It is hard to supercool water in your household freezer because there are far too many CCNs present in the air and in your tap water. As soon as that water hits 0°C, it freezes. But doubly distilled water (purified by boiling and condensing it twice), carefully handled, can be easily chilled to -10°C and stay liquid. For tiny water droplets in a cloud, the Kelvin effect also kicks in to help them supercool, so they readily supercool down to -40°C (-40°F: yes, °C and °F are the same at this temperature) without it freezing. Once the cloud temperature reaches -40°, however, the droplets spontaneously freeze. The moment that happens, there are now millions of tiny frozen ice crystals floating in the cloud, and these are some of the best CCNs available anywhere. They begin to nucleate other ice crystals like crazy and a cascade of nucleation kicks in.

Here is how the cascade works: clouds at temperatures below freezing consist of a lot of minute supercooled droplets of water, droplets too small to coalesce into raindrops and which therefore are doing little to make it rain or snow. They are just floating about waiting for something to happen. When the tops of such clouds are forced high enough that the top temperature dips below -40°C, then minute ice crystals start forming up there and these start filtering downward through the cloud of supercooled water droplets. The ice crystals start nucleating the waiting droplets as they float by, and the droplets freeze into minute snowflakes.

It is one other quirk of physics, however, that makes this something like a magic process. The quirk was set in place way back in the Big Bang when oxygen and hydrogen took on their atomic character, and it greatly amplifies the growth rate of the falling ice crystals. It turns out, due to the nature of the water molecule, the vapor pressure over liquid water (even when supercooled) is higher than the vapor pressure over ice at the same temperature. This is because the water molecules are more tightly bound in the ice solid than the liquid. The maximum difference between the two occurs at -12°C, but at -40°C the vapor pressure over water is still 30% higher than over the ice. All else being equal, then, a water droplet has to give up vapor molecules to an ice crystal nearby. It is like forced charitable giving. The ice will grow and the water droplet will shrink, and this transfer will happen to billions of droplets in the cloud. If the cloud is moist enough to sustain the process long enough, it will soon snow (Fig. 2-4). This atomic quirk is called the Wegener-Bergeron-Findeisen (WBF) process, taking its name from its three co-discoverers. Alfred Wegener is famous for originating the theory of continental drift and the prestigious Alfred Wegener Institute in Bremerhaven, Germany, is named for him. He died in Greenland on an expedition studying climate and snow, so he is also a hero to snow scientists (see *Alfred Wegener: Science, Exploration, and the Theory of Continental Drift*, by Mott T. Greene [Johns Hopkins University Press, 2015], 675 pages).

But the cloud crystal factory (Fig. 2-5) is not done with its work yet. Those first embryonic snow crystals are too small and light to fall out of the bottom of a cloud. The littlest updraft blows them back up, and they begin cycling up and down like a leaf caught in a whirlwind. Gravity pulls them down; updrafts send them back up. In each cycle, they

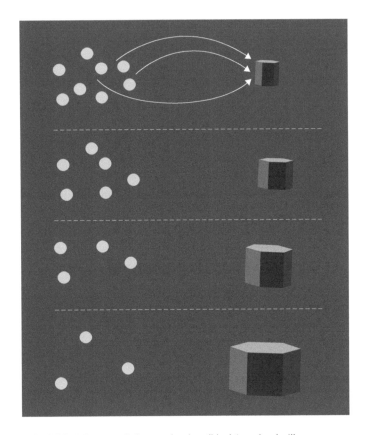

FIGURE 2-4. Supercooled water droplets (blue) in a cloud will preferentially lose water molecules (white arrows) to any nearby ice crystal due to the vapor pressure differences of water versus ice. In time (steps 1-2-3-4), there will be fewer water droplets and bigger ice crystals. Eventually, this cloud will start to snow.

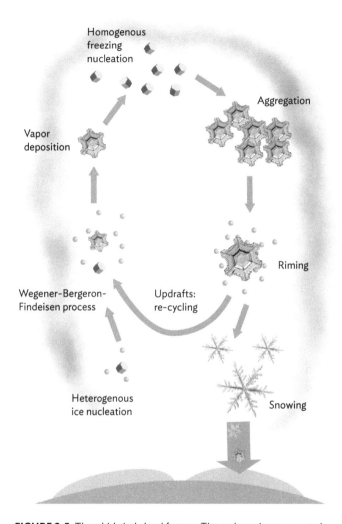

FIGURE 2-5. The whirlwind cloud factory. The embryonic snow crystals fall, grow, aggregate, rime, and grow until they are heavy enough to withstand the updrafts within the cloud and then they fall to Earth as snow.

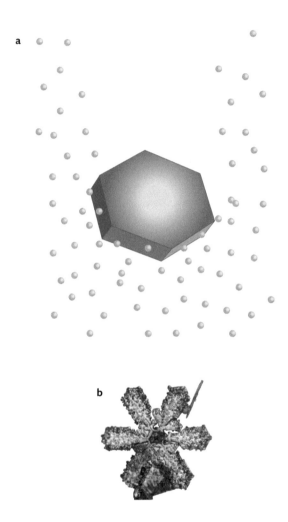

FIGURES 2-6a AND 2-6b. Riming takes place when an ice crystal falls through a zone in a cloud where there are liquid water droplets of substantial size. The crystal by now is large enough to have some cold content, so as it tumbles and falls, it smacks into these supercooled droplets and they freeze on contact, producing a speckled (or diseased) appearance to the snow crystal. (Photo by M. Sturm.)

grow bigger by water vapor deposition, also called **accretion**, by crashing into each other and sticking together (**aggregation**), and by **riming** (direct freeze-on of liquid drops: Figs. 2-6a and 2-6b). Eventually, gravity wins the battle and the enlarged snowflakes, now heavy enough to fall to Earth, drop into the open mouth of a running child.

Like an old cowhand, whose life history can be read in the scars on his hands and face, these snow crystals have features that tell us the tale of their cloud birth and what happened to them on their way down to Earth.

3

SNOWFLAKES

One day my mentor, the well-known Arctic snow scientist Carl Benson, handed me a book called *Snow Crystals: Natural and Artificial* by a Japanese researcher named Ukichiro Nakaya. I was blown away. The book is a marvel, combining both high art and great science, which individually are rare, and combined, exceedingly rare. Born in 1900 in Kaga, Japan, Nakaya studied physics, receiving a PhD in the 1920s. He was sent to Hokkaido University in 1930, where there was ample snow, but little physics equipment. So he began to study snowflakes, first classifying them by shape, then exploring how they grew and why they took on certain forms. He did this by learning how to grow snowflakes in a cold laboratory. It was exacting and at times frustrating work, but in time, by using rabbit hairs and other fibers as CCNs, he was able to replicate what fell out of the sky. He also became a superb photographer of both artificial and natural snowflakes, and it is these photographs that make his book a great work of art. Nakaya called snowflakes and crystals "letters sent from heaven," meaning they bore in their shape and form information about the clouds from which they had come. My mentor's copy of Nakaya's book, signed by the author himself, now sits in a reverential glass case in the library at my university. In Japan, a museum honors Nakaya and his likeness appears on a postage stamp.

We all love snowflakes and we have all heard that no two are alike, but any good snow observer ought to be able to answer these questions as well:

1. Why are snow crystals hexagonal?
2. How many types of snowflakes are there?
3. What do the type and size of snowflakes tell us?
4. Why are they symmetrical (or are they)?
5. Are any two snowflakes identical?

First, though, some terminology. A crystal is a particle whose molecules are arranged in a highly ordered repeating pattern, which gives rise to flat facets intersecting at sharp angles. In short, a recognizable geometric structure. Crystals look like gemstones on an engagement ring. Most published snowflake photographs are of snow crystals (not snowflakes) because these sharp-edge crystals are so cool. Snowflakes can be a single crystal, but the majority of what actually falls from the sky are multiple crystals interlocked and packed together, often irregular in form, frequently broken. These are not so pretty, and not nearly as often photographed or displayed on websites, as those single, symmetrical beauties.

3.1   THE ICE (OR WATER) MOLECULE

It is the structure of an ice molecule that creates the hexagonal symmetry seen in snow crystals, but surprisingly, the unit cell structure of ice is not a hexagon. It is a tetrahedron. Each oxygen atom forms the center of the tetrahedron with the hydrogen atoms making up the four tetrahedral corners (Fig. 3-1). Since the chemical formula for water is H_2O, implying there should only be two hydrogens in the structure, what is the catch? This paradox is explained by something called the *hydrogen bond* (H-bond). Due to their charge, the hydrogen atoms (each with a + charge) get shared between two neighboring oxygen atoms (with – charges), essentially vibrating back and forth. Linking a series of these hydrogen and oxygen unit cells together, it quickly becomes apparent that each oxygen can claim only two out of the four neighboring hydrogens associated with it, hence the chemical formula works. When a network of these tetragonal building blocks is connected, however, the hexagonal structure (Fig. 3-2) of ice becomes apparent. This form of ice is called Ice 1h; the *h* is for hexagonal. There are other forms of ice (including a cubic form) that form under enormous pressures, but most of the ice on Earth, including snow, is ice 1h and has hexagonal symmetry. In some ways, building snow crystals is like building with Lego blocks: a myriad of shapes can be built, but the shapes invariably reflect to some extent the fundamental shape of the blocks themselves.

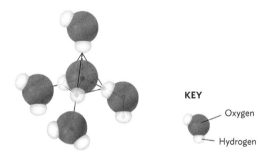

KEY

Oxygen

Hydrogen

FIGURE 3-1. The oxygen-hydrogen tetrahedron. The hydrogen atoms are shared by neighboring oxygen atoms, as indicated by the chemical formula for a water molecule (H_2O), so for the unit cell, each oxygen can only claim two of the four hydrogen atoms, shown in gray.

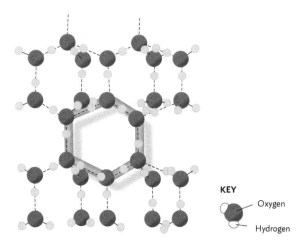

KEY
Oxygen
Hydrogen

FIGURE 3-2. When a series of water molecule tetrahedra (Fig. 3-1) are linked in a lattice, the hexagonal symmetry emerges in both horizontal and vertical planes (orange highlight), and this basic motif is apparent in virtually all falling snow crystals and several types of snow grains found in the snowpack.

3.2 BASIC ICE CRYSTAL FORMS

The basic motif of all snow crystals is the hexagonal prism (Fig. 3-3). It has two basal faces, the top and bottom, and six prism faces. There are only two growth modes. One is *c-axis growth*, in which case as the snow crystal grows by vapor deposition, it becomes long and thin, quite like a needle (see the Shimizu crystals in the sidebar, Figure S3-2 (p. 42), for an extreme version of snow needles). The other is *a-axis growth*, in which case a wide, flat, hexagonal plate develops. Virtually every snow crystal form can be produced by varying these two growth modes over time and combining them in various ways.

However, two other physical concepts are needed to understand the bewildering variety of snow crystal forms

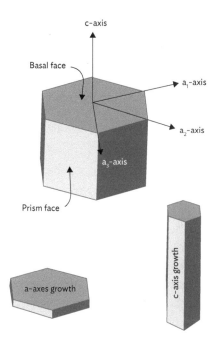

FIGURE 3-3. The basic motif of a snow crystal is a hexagonal prism that has one c-axis and three symmetrical a-axes. Crystal growth takes place along these axes, and can produce either hexagonal plates or hexagonal columns, depending on which growth mode dominates. Intricate forms arise from complex combinations of the two growth modes (plate and column) alternating back and forth as the crystal grows larger.

that can be observed. The first concerns the latent heat associated with vapor condensation. Snow crystals grow when water vapor molecules condense out on points, edges, and surfaces of an existing cold crystal, but this process releases considerable heat due to the change of state from a gas to a solid (the latent heat of vaporization). If the part of a crystal that is growing is unable to dissipate that heat, then at that place on the crystal, the temperature will begin to rise, and once it is warmer than nearby locations, it will cease to be a favorable (cold) location for condensation. Growth there will stop and sublimation (i.e., evaporation, but from a solid to gas) will start, actually leading to a loss of ice and a rounding of the crystal edges.

Car radiators and the cooling coils on refrigerators are designed to get rid of heat. They do this by having a lot of surface area in the form of fins and blades. Snow crystals are the same. They begin to develop branches, creating more and more surface area and thereby become more efficient at air cooling. For a snow crystal actively growing, forms like fins and branches will grow at the expense of flatter, more compact forms, and so branch tips will grow faster than nearby branch edges. This is why the dendritic (star) forms are so common in snow crystals.

The second concept concerns vapor delivery. The water vapor in a cloud has to make it to the crystal surface for growth to occur, and it is proportionately harder for the vapor to make it into deep concave recesses, or groins, then it is to make it to the tips of branches and fins. It is also just slightly harder to deliver vapor to the middle of a flat prism face than to the edge of the face. All things considered then, branch tips grow better than branch edges, which grow better than prism edges, which grow better than prism faces, all because of the ease of water vapor delivery. All this leads to the development of vapor-starved pockets in crystals, which once formed, cease to grow and often deepen. It is a self-reinforcing process. Because the crystal branches stick out into the supersaturated air farther than the edges of the crystal, they not only cool faster than their cloistered neighbors, but they also glom

| Points and face edges grow faster than the middle of faces. | Cavities and groins form, then become starved for vapor. | Branches develop; tips grow rapidly, branching helps dissipate latent heat. |

FIGURE 3-4. Hexagonal plate crystals will begin to grow preferentially at points while developing cavities and groins between the points (because corners and edges grow better than faces). This growth will produce stubby branches and a transition to sectored plates. If the branches continue to grow at the expense of the edges and faces, the cavities and groins will deepen and a stellar dendrite (right) will begin to develop. This is the form most people think of as a snowflake, though actually 120 other forms are now recognized by scientists Kikuchi, K., Kameda, T., Higuchi, K., & Yamashita, A. (2013). A global classification of snow crystals, ice crystals, and solid precipitation based on observations from middle latitudes to polar regions. *Atmospheric research,* 132, 460-472.. See also http:// snowcrystals.com. (Montage by Sturm; images from Ken Libbrecht. Used with permission.)

onto more of the available vapor supply. The faster they grow, the more they stick out, and the deeper and more inaccessible the pockets become. The result is that growing hexagonal plates (Fig. 3-4 left) turn into sectored plates, which branch out and become stellar dendrites (Fig. 3-4, right).

I rather like the way F. C. Frank, a British materials scientist knighted for his work, described this process in 1982. He called it the "lacunary catastrophe." A lacuna is an empty space or missing part. At the center of the crystal faces, where unfavorable thermodynamics (too much latent heat) and vapor starvation (hard to reach pockets) occur, holes and hollows develop. This will lead to the development of

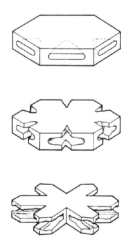

FIGURE 3-5. F. C. Frank's sketch of how the lacunary catastrophe (i.e., places on a snow crystal where it is too warm and/or not enough water vapor is available for that part of the crystal to grow) creates hollows and branched crystals.

(F. C. Frank, "Snow Crystals," *Contemporary Physics* 23, no. 1 (1982): 3–22.)

separate sectors in what were originally simple hexagonal prisms, and in time more and more ornate snow crystals will be produced (Fig. 3-5).

3.3   SNOW CRYSTAL CLASSIFICATION SYSTEMS

In 1936, Nakaya and a colleague, Sekido, classified snow crystals into 21 types. Sixteen years later, Nakaya expanded on his earlier work by identifying 42 snow crystal types. A decade after that, his colleagues Magono and Lee at Hokkaido University expanded the system to a full 80 types. Japanese scientists continue to lead this type of research, with Kikuchi and his coworkers in 2013 identifying 121 types of snow crystals.

Most of the new types have been found in the Arctic or Antarctic and are rare. One hundred and twenty-one is a bewildering number, probably far too many for most hobbyists to learn. Ken Libbrecht, the Cal-Tech physicist who has probably done more to promote snowflake watching than anyone else, suggests a simpler system (Fig. 3-6) with just 35 types. Of these, over 30 preserve the basic hexagonal symmetry shown in Figures 3-2, 3-3, 3-4, and 3-5 quite well.

Most people, even those living where it snows frequently, are unaware of the delightful range of snow crystal types that exist, yet it takes almost no equipment to make snow crystal observations. A magnifying glass or small hand lens, a parka or jacket made of dark material on which snowflakes can land, and a simple awareness of the snow, is enough to open the world of snowflake observing to an astute individual. Patience is useful as well, as during most storms, the type of snow crystals falling evolves (see next section) and several types of crystals may be observed in just a few hours. During a snowstorm, frequent checking of what is currently falling is often rewarded with some rarer forms, particularly as the snowstorm tails off. A small board covered with black velvet and a straw plucked from a broom for separating crystals on the board will make it easier to see what is falling, and if one is trying crystal photography, the straw can be used to isolate attractive individual crystals. Teasing the crystal onto a glass slide and spraying it with art fixative (like Krylon K01306) that has been chilled below freezing will produce acceptable rubberized casts of the crystals that can then be viewed indoors with lens or microscope.

Photography adds some complexity to snow crystal observing, but modern digital cameras and cell phones

Simple prisms	Solid columns	Sheaths	Scrolls on plates	Triangular forms
Hexagonal plates	Hollow columns	Cups	Columns on plates	12-Branched stars
Stellar plates	Bullet rosettes	Capped columns	Split plates and stars	Radiating plates
Sectored plates	Isolated bullets	Multiply capped columns	Skeletal forms	Radiating dendrites
Simple stars	Simple needles	Capped bullets	Twin columns	Irregulars
Stellar dendrites	Needle clusters	Double plates	Arrowhead twins	Rimed
Fernlike stellar dendrites	Crossed needles	Hollow plates	Crossed plates	Graupel

FIGURE 3-6. Thirty-five basic types of snowflakes derived from the classification of Magono and Lee (C. Magono and C. W. Lee, "Meteorological Classification of Natural Snow Crystals," *Journal of the Faculty of Science,* Hokkaido University 2, no. 4 1966) as simplified by Ken Libbrecht of Caltech. (Ken Libbrecht images. Used with permission.)

FIGURE 3-7. This stellar dendrite fell on a cold plastic bag while we were making snow depth measurements, and I snapped a picture using a compact Olympus TG-5 digital camera that has a microscope mode (crystal diameter is 1 mm [0.04"]). Cell phone digital microscopes are also now available for less than $100 that can take equally fine images of snow crystals. (Photo by Matthew Sturm.)

have made even this aspect of snow crystal observing easier. Compact digital cameras with macro- or even micro-focusing capability produce excellent images. Several of the newer cameras (I use an Olympus TG-5) have image stacking capability, which alleviates many of the older focusing and depth-of-field issues (Fig. 3-7). A simple stand with a short working focus distance, just a few centimeters, is needed to steady the camera and get the crystals in sharp focus. My colleague Kelly Elder has a design for a simple camera stand made of Plexiglas that takes just an hour to build. Cell phone microscopes with focusing stands now sell for less than $100 and can also produce quite acceptable images as well (e.g., AmScope USB microscope UBW500X02MP).

3.4   WHY SO MANY TYPES OF CRYSTALS?

The secret as to why there are so many types of snow crystals can be found in the clouds (Fig. 2-5). A typical winter cloud is colder at the top than the bottom exhibiting a wide range of water vapor saturations as a function of height, and these conditions are constantly changing over time. The cloud also has wildly varying up- and down-drafts, these too varying rapidly. An incipient snow crystal (as described in Chapter 2) goes for a wild ride in such a cloud. Initially, being so light and small, it has a slow fall rate and the updrafts in the cloud keep sending it back aloft. Eventually, it will grow heavy enough that it begins to settle and fall. As the snowflake falls it will travel downward through cloud layers that have very different temperatures and vapor saturation conditions than at the level where it originated. The crystal growth rate and habit (c-axis versus a-axes growth) will keep changing as the crystal makes this downward journey. By the time it lands on the dark sleeve of an observer, it will have gone through multiple distinct growth cycles, and its form will reflect the journey. This is what Nakaya meant by snow crystals being "letters sent from heaven."

Conditions in a cloud can be best represented by two variables, temperature and supersaturation (i.e., the amount of moisture in the air). For any pair of these variables, there is a distinct crystal type associated. The transitions between types are not gradational, but rather quite abrupt, a phenomenon that still escapes complete scientific explanation. Plotting crystal types on a diagram with these two variables is known as a Nakaya diagram (Fig. 3-8), and it lies at the heart of snowflake understanding. A snow crystal falling through a cloud traces a circuitous pathway through the conditions represented on this diagram, but since we rarely have tempera-

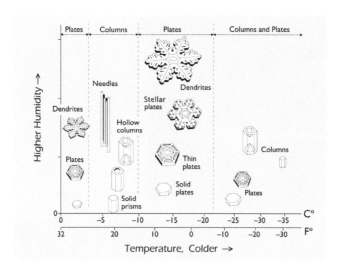

FIGURE 3-8. A Nakaya diagram. The snow crystal type is controlled by temperature and supersaturation (cloud humidity), with abrupt transitions in crystals occurring at -3°C, -10°C, and -21°C (27°F, 14°F, and -6°F). (Redrawn by Ken Libbrecht.)

ture and supersaturation measurements from the cloud, we have to infer the path from the crystal form. There are nearly an infinite number of possible paths, too. For example, the tiny hexagonal plate in Figure 3-9a (crystal center) began to turn into a sector plate as it grew, suggesting the temperature where the crystal was growing was -15°C and the humidity was changing from mildly to heavily supersaturated. At some later time, the crystal reverted to the hexagonal plate form as its edges grew out, suggesting the supersaturation dropped, but then it must have risen once again, causing the crystal to begin to branch. In this way the form of the snow crystal suggests its circuitous pathway up and across the Nakaya diagram, with the end of its journey on our sleeve.

FIGURES 3-9a AND 3-9b. A sectored plate that began as a hexagonal plate (1); started to branch (2) and (3); reverted to hexagonal plate (4); then began segmenting into a sectored plate (5). This particular snowflake photo was taken by Wilson A. Bentley (Fig. 3-9b), a Vermont farmer who in 1885 began the first-ever series of snow crystal photographs. By the end of his life (in 1931) he had taken over 5,000 images of snow crystals and was world famous for his work. Many American children learn of Bentley (at work here with his bellows camera) from the book *Snowflake Bentley*, and his wonderful snowflake images can be used by anyone for noncommercial activities (http://snowflakebentley.com/WBsnowflakes.htm).

(Source: Jacqueline Briggs Martin (author) and Mary Azarian (illustrator), Snowflake Bentley [New York: Houghton Mifflin: 1998].)

3.5 SYMMETRY

The symmetry of snowflakes is initially due to the structure of the water molecule, which results in beautifully symmetrical hexagonal micro-crystals. As these grow larger, because they are *tumbling* and *rotating* as they fall through the cloud, each edge or branch is exposed to much the same growth conditions as the other ones. All the branches tend to grow the same, producing the symmetry. In reality, though, non-symmetrical snowflakes are quite common, more often the rule, but as they are not as photogenic as their symmetrical cousins, the latter get all the attention.

3.6 ANY TWO ALIKE?

The long-running argument as to whether any two snowflakes are identical is really more semantic than real. At the molecular level, no two snowflakes are likely to ever be the same. Consider how many water molecules are in the snow crystal shown in Figure 3-7. It weighs less than a hundredth of a gram, but still, it is made up of over 10^{20} molecules (that is nearly a billion trillion). Due to lattice imperfections, no two snow crystals could ever be arranged exactly the same at this molecular level. More practically, the question might be whether any two snowflakes *look* the same. The answer to this question is certainly *yes*. Particularly for simple, unbranched crystals and diamond dust (see "Diamond Dust" sidebar [p.41]), it is highly likely that two snowflakes will look so similar as to be identical to a human observer.

3.7 FALL SPEEDS AND SNOWFALL INTENSITY

Sadly, at least 98% of all snowflakes are imperfect, a number readily confirmed once a person gets involved in snowflake observing and photography. Good specimens can take time to find. Irregular, broken, asymmetrical, partially melted, or rimed crystals tend to be the rule. The best specimens are usually found when fall speeds are low, when there is little wind to break up the crystals, and when the snowfall intensity is low as well. This last condition means that single crystals rather than crystal aggregates are likely to be falling. These are just opposite the conditions that favor the build up of a good, thick snowpack, the subject of the next chapter, and the ardent desire of skiers. There is a wide range of sizes and masses of falling snowflakes, but a surprisingly narrow range of fall speeds (Fig. 3-10), which generally run 1 to 2 meters per second (3.2 to 6.4 feet/second). At 1 meter per second, catching a snowflake on the tongue is easy; at 2 m/s it gets harder. Hail, however, can come rocketing down from a cloud, as evidenced by these record hailstones (Fig. 3-11) that came barreling down at a lethal 50 meters per second (111 mph).

Snowfall intensity is more difficult to measure than fall speed. NOAA reports 24-hour snow *depth* totals, a value that favors fluffy low-density snow even though a greater depth value may actually represent less snow water equivalent (SWE) than a nearby location with a lower 24-hour depth number. Nevertheless, because the water equivalent of snow over 24 hours is not widely measured, snow depth is what is used for the record books, and setting a snow record is highly coveted. The rules for setting the record are also strict: depth can only be measured four times in a 24-hour period, then added together, and the method of measurement must be NOAA approved.

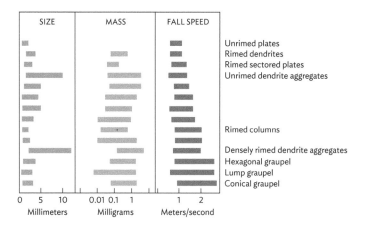

FIGURE 3-10. While the size of snowflakes ranges over a factor of 10X, and the mass over a factor of 200X, the fall speed of snowflakes ranges only from about 1 to 2 meters per second, with aggregate and rimed snowflakes (Figs. 2-6a and 2-6b) falling faster than unrimed and single-crystal snowflakes. (Redrawn from Locatelli, J. D., & Hobbs, P. V. (1974). Fall speeds and masses of solid precipitation particles. *Journal of Geophysical Research*, 79(15), 2185-2197..)

As Figure 3-12 shows, the *official* 24-hour record (78 inches or 198 cm) is held by my state, Alaska. Next highest (75.8 in.) comes from high in the Colorado Rockies, but both New York (with fierce lake effect snowfall near Buffalo) and California (with vast rivers of moist air driving into the Sierras from the Pacific Ocean) continue to vie for the prize. In 1952, a massive 84 inches (213 cm) of snow fell at Crestview, California, in one 24-hour period, but as this was not sanctioned as a record, it does not officially stand. The Alaska record almost certainly has been exceeded somewhere else in the U.S., and the record, set in 1963, has been questioned (see http://us-climate.blog-spot.com/2014/05/mile-camp-47-snowfall-record.html), but subsequent investigations indicate the measured value is plausible and so it still reigns as the national champ.

FIGURE 3-11. Record hail that fell on July 23, 2010, in Vivian, South Dakota, at more than 50 meters/second (111 mph).

(NOAA, https://www.weather.gov/abr/vivianhailstone.)

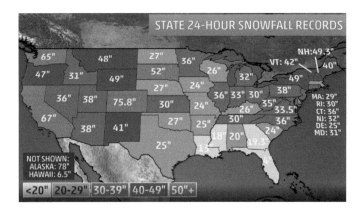

FIGURE 3-12. State 24-hour snowfall records in inches. Multiple by 2.54 to get centimeters. (NCDC-NOAA.)

DIAMOND DUST

There is a special type of snow crystal that forms relatively close to the ground under clear skies at very low (-40°C/-40°F) temperatures called *diamond dust*. These crystals are minute, only 50 to 250 μm (0.05 to 0.25 mm, or a few thousandths of an inch) in length; five of these crystals could sit on the point of a pencil. Diamond dust forms during clear weather, with no clouds in the sky. It can form at night or in the daytime, but it takes its name from its appearance by day: the sky seems to be filled with millions of tiny particles sparkling in the sunlight. The sparkling arises from the fact that the dust settles out of the air slowly because the limited weight of the crystals is balanced by their air resistance in falling. The slowly falling basal and prism faces of the crystals form perfect little prisms for reflecting and refracting sunlight.

Diamond dust crystals (Fig. S3-1) can be short stubby columns, long thin columns, hexagonal plates, or needle-like crystals called *Shimizus* (Fig. S3-2). The type of crystal that forms is primarily a function of the temperature, as the Nakaya diagram suggests (Fig. 3-8). Unlike snowflakes, which are larger and have grown longer and in a

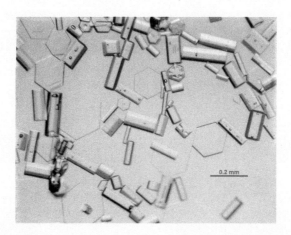

FIGURE S3-1. Diamond dust crystals photographed by Walt Tape. Note that, indeed, two of these snow crystals can look very much alike. (Photo by Walt Tape. Used with permission.)

41

more complex environment, diamond dust crystals are simple and elegant. Unfortunately, they are so small it is difficult to view them without a good microscope.

The interaction of diamond dust and sunlight gives rise to sun pillars, parhelia (sun dogs), halos, and other spectacular lighting phenomena (Fig. S3-3). These phenomena are most often seen in the polar regions, but they can be seen in lower latitudes at times. In fact, small rainbow spots seen near high cirrus (ice) clouds in the tropics are actually formed by diamond dust.

Scientists now mostly understand how arcs, halos, and sun dogs form. The diamond dust has two basic forms: flat hexagonal plates and long hexagonal columns. The plates and columns settle through the air in different ways (S3-4) due to air resistance. The plates settle like leaves with their hexagonal surfaces facing up or down. The columns have their maximum air resistance on their long sides, so they settle the other way, with their hexagonal surfaces vertical. Within a cloud of diamond dust, most of the plates or columns will be lined up to within a degree or two of the orientations shown in Figure S3-4. The result is that the dust makes up a giant prism, composed of millions of individual crystals, all nicely aligned, which refracts sunlight in a predictable way (Fig. S3-5). The process is not all that different from hanging a crystal prism in a sunny window and having it produce rainbow spots on the walls.

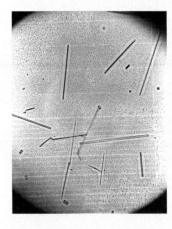

FIGURE S3-2. Not common, but one exotic type of diamond dust are these long, thin Shimizu crystals, the ultimate in c-axis growth. These Shimizu crystals were photographed at the South Pole (Walden, V. P., Warren, S. G., & Tuttle, E. (2003). Atmospheric ice crystals over the Antarctic Plateau in winter. *Journal of Applied Meteorology, 42*(10), 1391-1405). The distance between the upper white scribed lines is 0.25 mm (10 thousandths of an inch).

(Photo by Steve Warren. Used with permission.)

FIGURE S3-3. Sundogs, arcs, and halos caused by diamond dust in the air.
(Photo by Walt Tape. Used with permission.)

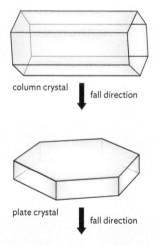

column crystal ↓ fall direction

plate crystal ↓ fall direction

FIGURE S3-4. The orientation of plate and column crystals as they fall through the atmosphere.
(Matthew Sturm)

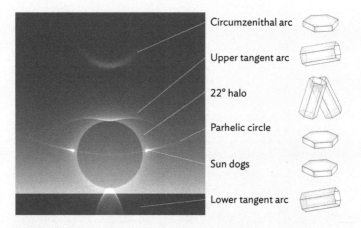

Circumzenithal arc

Upper tangent arc

22° halo

Parhelic circle

Sun dogs

Lower tangent arc

FIGURE S3-5. The relationship between the type of diamond dust crystal and the atmospheric display. (Photo by Walt Tape [left]. Graphics by Matthew Sturm [right].)

4

SNOW LAYERS AND WEATHER

Rafters going down the Grand Canyon (Fig. 4-1) learn about sedimentary and metamorphic rock layers. These layers get their character (orientation, grain size, color, sorting, etc.) and their classification (mudstone, sandstone, limestone, or dolomite) from the environmental conditions under which they were deposited. Their thickness is a product of deposition rate and time. In some cases, there is a long hiatus between the deposition of one layer and the next (called an unconformity), so the stratigraphic column has time gaps. Once deposited, the layers weather and change, compacting and bonding due to chemical and thermal processes as well as groundwater seepage. Subsequent geologic events may deposit more layers or erode old ones. Geologists call these collective processes *diagenesis*. Snowpack layers develop in virtually the same way, though instead of taking millions of years, the processes take days to months.

4.1   SNOW LAYER CHARACTERISTICS
Snow layers are characterized by several key measures including:
- Thickness
- Density

FIGURE 4-1. The layered stratigraphy of the Grand Canyon and the layers of a snow cover have much in common, drawing their character from the conditions that prevailed when the sediments (sand or snow) were deposited. (Photo by Betsy Sturm. Used with permission.)

- Porosity
- Hardness
- Grain size and shape
- Temperature
- Stability (shear strength)

The thickness of a new snow layer is a product of how much snow fell in a storm, and how much that snow compacted after it was deposited but before it was measured. Those record snowfall events discussed in Chapter 3 that dropped between 1 and 1.5 meters (40 to 60 inches) of snow in 24 hours produced very thick but also very fluffy layers. Within a few days these layers would have settled considerably, with an initial 1.5 m (60-inch) layer reduced to 0.75 m (30 inches) or less over that interval. At the same time, the layer density (defined as the mass of snow in a unit volume) would have increased proportionately. Despite the substantial compaction, we would still be able to distinguish such a layer in a snow pit. It would have a different density and different grain characteristics than the layers below and above, and we would almost certainly still be able to detect the upper and lower layer boundaries.

Density is one of the most important layer attributes we can measure. It is measured by taking a known volume of snow using a sampler, then weighing the sample. A 1-liter volume (1,000 cm^3 of water) weighs 1,000 grams, so its density is 1,000 g divided by 1,000 cm^3, or 1 g cm^{-3} (alternately, 1,000 kg/m^3). We can express density as a unitless number called **specific gravity**, which is the ratio of the volume of one material to the same volume of water. A 1-liter volume of bubble-free ice weighs 917 g, and has a specific gravity of 0.917. The value being less than 1 explains why ice floats in

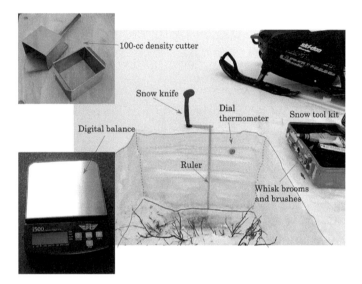

FIGURE 4-2. A typical snow pit with some of the tools used to measure snow layers. Many of the tools can be improvised from kitchen utensils or bought in a hardware store. A number of different designs for density cutters (upper left) are available; all work on the same principle of sampling a known volume and then weighing the mass in that volume.

water. A set-up for measuring density in the field is shown in Figure 4-2. The steel cutter in the inset on the upper left is pushed into a snow layer exposed in the pit wall and a precise volume of snow cut off, then weighed on the digital balance on the lower left. The density is computed from the mass reported by the balance divided by the volume of the cutter.

The air contained in the snow is what makes it less dense than ice. The porosity (φ) tells us (in fractional form) just how much air there is. Porosity can be computed from the snow density (ρ_{snow}) by a simple formula:

$$\varphi \left(porosity \right) = 1 - \left\{ \frac{\rho_{snow}}{\rho_{ice}} \right\}$$

where ρ_{ice} is 0.917 g/cm³. As Table 4-1 indicates, the porosity of newly fallen snow layers is extremely high, nearly 90%. For any snow we are likely to encounter under normal winter conditions, the air content will be more than 50%. Given this high percentage of air, we might expect layers to settle quickly, expelling the air as they become denser due to the pull of gravity, and indeed to a large extent this is what happens. As more new snow falls on top of a layer, the settlement rate accelerates. Temperatures near freezing, strong sunshine, and rain also speed up the settlement process. But as the layer gets denser, it also begins to resist settlement. This creates the elegant settlement curves called out by number in Figure 4-3. Eventually, settling layers reach a quasi-stable density controlled by the type of grains in the layer, the temperature, and the weight over the overlying layers.

TABLE 4-1. Density of newly fallen snow layers by type

	Snowflake type	Density (g/cm³)	Porosity (air fraction) (φ)
Unrimed	dendrites	0.056	0.94
	needles	0.060	0.93
	plates	0.087	0.91
	irregular	0.095	0.90
Rimed	needles + dendrites	0.093	0.90
	dendrites	0.109	0.88
	plates	0.143	0.84
Other	wet snow	0.300	0.67

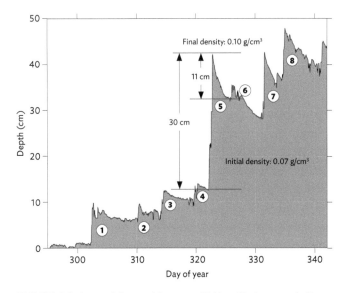

FIGURE 4-3. A snowfall record from near Valdez, Alaska, recorded by a sonic sounder. Eight separate snowfalls are shown during this forty-day period, the biggest a 30-cm dump (12 inches). Following each snowfall, the depth decreases in an elegant settlement curve, fast at first then more slowly as the new snow gets denser. Over three days the 30-cm dump settled 37% (11 cm or 4.3″) while the snow density increased a commensurate amount.

Since most snow is composed of at least 50% air, one might wonder how a buried avalanche victim can die from lack of air. It is because porosity and permeability are not the same thing. Porosity, as defined above, is simply how much air space is in a material, not whether those spaces connect. Permeability, however, describes the ability of air to flow through a material, and is a strong function of pore space connectedness. Two pieces of foam, one open cell and the other closed cell, might have the same porosity, but the closed cell foam will not allow air to move through its pores, while the open cell foam will. For a victim buried by an avalanche, high permeability matters greatly, but due to the victim's own body heat and moist breath, an impermeable icy barrier is soon created near their face, which cuts off the flow of air in the snow. Unless the victim is rescued quickly, he or she can suffocate.

So when the snow is not new, what is its density? I am frequently teased by colleagues about the "rule of point three": we measure hundreds of snow layer densities every field season, yet the average always turns out to be around 0.3 g/cm^3, a number we could just as easily estimate as measure. Strangely, it turns out the thermal properties of snow layers also converge around 0.3, though the units are different. But that said, it is worth having a general notion of what density to expect for a certain type of snow (Table 4-2). Other than snow layers found on a glacier that have survived through an entire year (called firn), the densest snow layers are invariably produced by the wind because it pulverizes snow and creates small irregular grains that pack together well. There is a theoretical upper limit to wind slab densities: it lies between 0.5 to 0.6 g/cm^3 and is related to the density achieved through the simple packing of ice spheres (or ice marbles). Glacier firn, which has been compressed and wetted for over a year, can be denser than wind slab because the snow grains will have deformed and the snowpack will have been infused with meltwater that filled pore spaces. Firn, with a density of 0.7 g/cm^3 or higher, looks and feels like ice, but can be distinguished from ice by simply seeing if one can blow into the material. For ice, which is impermeable, that isn't possible, but for firn it still is possible because the pore spaces are still connected despite the densification.

TABLE 4-2. Density (g/cm³) of typical snow layers

Type of snow layer	Density	Notes
New "wild" snow	0.05	The powder every skier dreams of
New snow	0.06-0.10	
Recent snow	0.10-0.15	Snow that fell 1 to 7 days ago
All snow	0.3	Based on over 25,000 measurements
Depth hoar	0.25	
Cascade "concrete"	0.4	Wet, maritime snow
Moderate wind slab	0.4	
Hard wind slab	0.5	
Corn snow	0.4	Spring snow that has been through melt-freeze cycles
Firn	0.55	Snow on a glacier more than 1 year old; still permeable
Simple cubic packing of ice spheres	0.56	
Ice layers (bubbly)	0.7-0.8	Pore close off in snow
Ice	0.917	

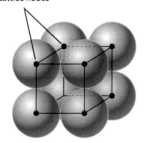

Lattice nodes

Simple cubic lattice cell

4.2 OBSERVING SNOW LAYERS

Thickness, density, and porosity alone are not enough to define and describe a snow layer. Grain size and shape, the temperature of the grains, and the strength of the layer due to bonding all contribute to its layer characteristics and the properties it will exhibit that might be important. For example, whether the layer will support the weight of a wolf running over the snow is a function of the grain size and grain bonding (the ice links between grains) as well as the layer density. Unlike measuring density, though, learning how to observe these additional attributes takes time. This is why snow pit observations are a craft as well as a science (Figs. 4-4a and 4-4b).

There are numerous online and published resources, recipes, forms, and instructions for digging and measuring layers in a snow pit, and many of these are good. Most are designed with a specific task or goal in mind, like making a snow stability assessment. I would contend, however, that if observing snowflakes can be done for fun and interest, so too can reading the story of the snow and winter in a pit. I practice what I preach: even on fun ski tours, with the nearest avalanche danger miles away, I often dig a pit just to see "what's there," and to do that I carry a minimum of gear: a shovel, a hand lens, and a ruler. Each observer will have their own reasons and needs for looking at snow layers and will need to develop their own protocols to suit their needs. Here, instead of a recipe, I provide some simple guidance based on digging and measuring thousands of snow pits. Adopt those parts that prove useful and discard the rest.

But before creating your own protocol, I highly recommend checking out three handbooks that lay out methods of observing and recording data about snow layers. The first

FIGURE 4-4a. A 150-cm-deep (60") snowpack near Nome, Alaska, in which we have excavated two pits to create a thin snow wall. We have then covered one side to create backlighting, which nicely exposes the layers. There are at least fifteen here, though some may be post-depositional icy layers, the result of water percolation down through the pack.

FIGURE 4-4b. Other than a snow shovel, only a few tools are needed to make layer observations: a folding ruler, a whisk broom, a gridded comparator card, a large kitchen spatula, and a microscope or good hand lens.
(Public domain.)

is *The International Classification for Seasonal Snow on the Ground (2009)*, and it can be downloaded from the UNES-DOC digital library.[1] Personally, I like the older *International Classification for Snow on the Ground (1990)* because it is shorter and simpler yet still has most of the same information in it. The older classification can still be obtained here.[2] Similar information on snow layers can be found in *Snow, Weather, and Avalanche Guidelines*, which can be purchased from the American Avalanche Association.[3] In addition to snow layer measurements, this book includes extensive material related to snow stability. But even without these guides, the sidebar "Making Snow Pit Observations" found later in this chapter provides a bare-bones checklist for digging a pit and measuring the layers. It is a good starting point for developing your own personal protocol.

While guides and tools are helpful in observing snow layers, I believe the observer's mindset is even more important. Unlike the layers of the Grand Canyon, snow layers are all white. Visual differences can be subtle, and lighting is crucial in seeing these. While the standard avalanche pit protocol is to dig a pit facing the sun in order to shade the pit face and retard solar melting, that shading can make seeing the snow grains and layers more difficult. So I often cut a pit with the sun on the face for that reason. Do what works for you.

---

[1] https://unesdoc.unesco.org/ark:/48223/pf0000186462
[2] https://www.google.com/url?sa=t&rct=j&q=&esrc=s&source=web&c-d=&cad=rja&uact=8&ved=2ahUKEwiozbiQ3a_sAhXWup-4KHWiEA4kQFjAAegQIBxAC&url=http%3A%2F%2Fhome.chpc.utah.edu%2F~u0035056%2Foldclass%2F3000%2FSeasonal_Snow.pdf&us-g=AOvVaw3quBUdJMCK4_qfonb9qj-a
[3] https://www.americanavalancheassociation.org/swag

Here are a few helpful rules:

- ***Dig a wide snow pit*** (Figs. 4-5a). Most snow layers pinch, swell, and change laterally. Exposing these lateral variations will clarify what the layer is really like and why, and will reveal far more than can be seen in a narrow, crowded pit.

FIGURE 4-5a. This is a wider snow pit than would generally be dug, but the width nicely exposes the gentle undulations of the layers. An FM-CW radar that can "see" snow stratigraphy is mounted on the snowmobile in the background and the results will be compared to the visual observations on the pit face. Note how clean and sharp the pit walls are. This too helps an observer see what is happening in the snow.

- ***Wear lots of clothes*** and stay warm. Looking at snow layers is inherently cold work because you are not moving, and cold observers tend to rush and miss clues and interesting features.

- ***Get close***. If the snow is less than about 150 cm deep, that means kneeling or lying down so your eyes are at the level of the snow layers and close. Bring a pad to kneel on.

- ***Allow time.*** When in production mode during science campaigns, we have to knock out a pit sometimes in as little as fifteen minutes, but when learning, several hours will be needed. Speed comes with practice.

- **Engage all senses.** Even smell is useful. In spring, the plants under the snow begin to photosynthesize and they smell wonderful.

- **Feel and manipulate** the snow and listen as you do so. A cheap palette knife (Fig. 4-5b) works well, and the chatter of the knife as it slides down the face suggests differences in grain size and bond strength.

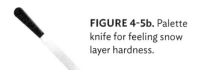

FIGURE 4-5b. Palette knife for feeling snow layer hardness.

- **Use markers** (like wooden coffee stirrers) to mark layer transitions so you don't have to hold too much information in your head. That will keep your head clear to absorb what your senses are telling you.

Lastly, even if you are drawing-challenged, before recording any numbers or words, try to **sketch** what you see (Fig. 4-6). Many of the characteristics of snow layers do not lend themselves easily to recording using just numbers or words, and while photographing a snow pit wall can be useful, the camera rarely captures everything the human eye can see.

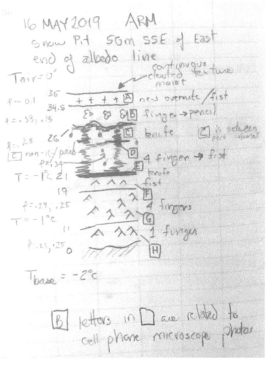

New snow   Wind slab   Depth hoar   Chains of depth hoar

Fine-grained   Melt grains   Ice lenses and percolation columns

FIGURE 4-6. A field sketch of the layers in a snow pit measured near Utqiagvik (Barrow), Alaska. Like any good set of notes, it includes when and where it was done. The layer density measurements are on the left side, denoted by the Greek letter rho, and the hand-hardness measurements are on the right (see the sidebar "Making Snow Pit Observations" for explanation [p. 66]). The air (T_{air}) and snow-ground interface temperatures (T_{base}) were recorded to provide information on the temperature gradient across the snow, though in this case the temperature of individual snow layers was not measured. The letters (A, B, C, etc.) key the sketch to grain micro-photographs we took using a cell phone microscope. The symbols used in the diagram (key at the bottom) are explained in more detail in Chapter 5.

4.3 WEATHER EFFECTS ON LAYERS

The diagenesis of snow layers takes place because gravity, wind, liquid water, and thermal processes begin to change the nature of the layers once they are deposited. Collectively, these can be called the effects of weather and time on the snow. Perhaps the easiest way to understand their impact is to examine a real case study from a place called Tsaina River, Alaska. It is located in the Chugach Mountains on the inland (drier) side of that range, a place where it snows a lot. In fact, the location is less than 10 miles (16 km) from where the U.S. 24-hour snow record was set (Chapter 3). Throughout the winter moist weather systems spinning off of the Gulf of Alaska push over Thompson Pass and dump lots of snow at Tsaina River, though occasionally a short rain-on-snow event will occur there as well. As a result of all that fluffy snow, the place is popular for off-piste helicopter skiing. The weather record in Figure 4-7 comes from a set of instruments mounted on a tower located in a spruce stand near the Tsaina River Lodge.

The records show that as winter began, air and ground temperatures began to drop, and by November the ground had frozen (its temperature drops below 0°C). Any snow that fell at this time would have landed on cold ground and would not have undergone basal melting. Initially, there was little wind, but in mid-November the wind rose to 9 m/s (about 18 mph). This is enough to drift snow (it takes 5 to 6 m/s to do this), so any snow that was in place at that time was likely to have undergone drifting. The air temperature kept dropping through December, reaching a chilly minimum of -30°C (-22°F) in the middle of the month. This set up a strong difference in temperature between the bottom

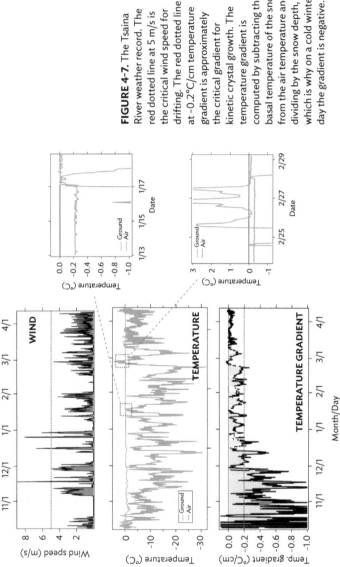

FIGURE 4-7. The Tsaina River weather record. The red dotted line at 5 m/s is the critical wind speed for drifting. The red dotted line at −0.2°C/cm temperature gradient is approximately the critical gradient for kinetic crystal growth. The temperature gradient is computed by subtracting the basal temperature of the snow from the air temperature and dividing by the snow depth, which is why on a cold winter day the gradient is negative.

and the top of the snowpack (this is called a ***temperature gradient***), with low temperatures at the top and relatively high temperatures at the bottom. Thermal gradients like this are important because they drive water vapor from warm to cold and that results in snow grain metamorphism (see Chapter 5). When the magnitude of the gradient exceeds about -0.2°C/cm (red dotted line in Figure 4-7 bottom panel), kinetic grain growth kicks in. This is a rapid type of growth that produces large hollow and ornate crystals called ***depth hoar***. As the gradient in parts of October and November were five times stronger than needed for kinetic growth, depth hoar development must have been rapid. By January, with the total snow depth increasing, the temperature gradient began to moderate (since it is the drop in temperature across the snow divided by the depth) and soon ceased to exceed the critical magnitude. A last cold snap in February reinitiated kinetic growth conditions, but briefly, and by late February midday temperatures were rising above freezing and causing surface melting (red dotted squares).

The two close-up graphs focus on the two melt events that occurred during the winter, and they were of very different character. In late February three midday spikes of +3°C temperatures were recorded during a period of sunny weather. It was warm enough these days to produce surface melting, but the amount of melt water would have been quite limited. No disturbance of the temperature at the base of the snow was recorded until late on the third day when a bit of water percolating downward finally arrived at the base. We can tell when the water got to the ground because the temperature there jumped up to 0°C in the space of minutes. The other melt event occurred on January 17, and it happened much faster. The air temperature rose abruptly in that event, but never above freezing, yet within two hours of that rise, the temperature at the base of the snow had spiked up 0.2°C and was pegged at zero degrees for the next several days. The only way that could have happened was a rain-on-snow event, an event confirmed by the records from the Valdez weather service. Rain water on the snow surface quickly percolated downward and pooled at the base of the snow, creating an ice-water bath that held the temperature precisely at 0°C until all the water finally froze.

Now let's add the snow depth record for that winter (Fig. 4-8), recorded using a sonic sounder similar to the one that reproduced Figure 4-3. The wind and rain events have been labeled, along with those periods when the temperature gradient was super-critical. These would have been when the snow was metamorphosing into depth hoar. The double wind event at the end of December does something odd to the depth record: the depth abruptly increases, then wind removes snow and it decreases. This is the signal a snow dune makes as it moves under, then past, a sonic sounder. The two liquid water events impact more than just the snow surface when they occur, with water percolating down through the pack and spreading out at layer interfaces and at the base of the snow. That ice, once in place, remains for the rest of the winter.

The actual snow layer stratigraphy as recorded in the field near the end of winter matches the weather-driven inferred stratigraphy reasonably well. The weather-derived layers tend to overestimate the thickness of depth hoar and wind slab layers, and, of course, provide only a rough guess as to where the icy layers and lenses formed internally, since that is a complex interaction of water and snow. Still, the weather-to-layer relationships don't do too badly, and illus-

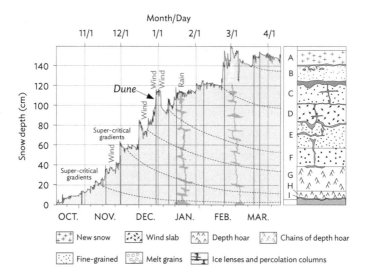

FIGURE 4-8. The Tsaina River snow depth (left) and the stratigraphy measured in March (right). The stratigraphic symbols used are the same as in Figure 4-6, but here the size of the dots suggest the snow grain size for fine-grained snow. As in Figure 4-6, the layers have been assigned letters to help identify them. Layer I, a basal ice layer, was deposited when rain water (January 17) and surface melt water (February 26–28) percolated to the base of the snowpack. The cold weather and thin snow of October and November produced depth hoar layers G and H. The wind slab (Layer D) was probably produced by the drift event in mid-December. The icy layers and percolation columns (dark gray symbols) throughout the pack were produced by the melt and rain-on-snow events after the deposition of many of the snow layers.

trate the general principles by which weather and time produce snow layer characteristics.

Four final points in pulling the story of winter out of snow layers:

1. It matters little by the end of the winter the type of snowflakes that make up a layer. They are long gone. In fact, as we will see in the next chapter, they are often "gone" in as little as thirty-six hours after deposition.

2. The composite of all the layers, when we can read them well, tells us the story of the winter: the cold snaps, the warm spells, the wind, and the rain. The layer measurements tie that information to the physical, thermal, and optical properties of the snow (Chapter 6), and it is these properties by which the snow impacts our ecosystems, animals, and our lives.

3. Layer order matters. Consider Figure 4-9. When we invert this simple two-layer snow system, we can go from an unstable configuration (for avalanches) to a stable one. Similar rules hold for water percolation and heat transfer, as well as and many other aspects related to snow layering.

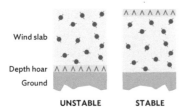

Wind slab

Depth hoar

Ground

UNSTABLE STABLE

FIGURE 4-9. The ordering of layers matters for many reasons, including avalanche stability, water percolation, and the ability of caribou to browse.

4. Remember: all geologic layers, including both snow and those in the Grand Canyon, vary as we move to a new location, because the depositional environment changes with macro and micro factors. In geology, these are called *facies changes* and they (Figure 4-10) can result in considerable differences in the snow layers.

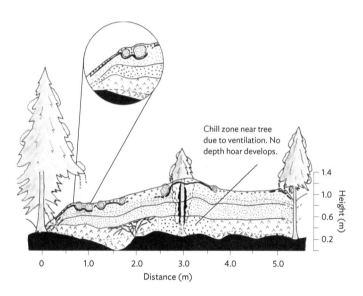

FIGURE 4-10. Snow layers in a forest vary with proximity to trees of different size, exhibiting complex facies changes due to changes in depositional and post-depositional conditions.

MAKING SNOW PIT OBSERVATIONS

1. Dig a snow pit (Fig. S4-1 [p.66]). Pile the snow out of your way, but where it is easily shoveled back in the pit. Sharp clean pit walls will help the observations. In locations where others are out in the snow, always backfill the pit when finished to prevent accidents. It is a good idea in general as the thermal disturbance to the snow will be minimized as well.

2. Once the pit is dug, arrange your tools adjacent to the pit in a way that they are near-to-hand for efficient measurements. Shelves can be dug in the pit side walls to hold some of the equipment. Ski poles can be stuck into the wall to provide handy hangers.

3. Smooth the pit face with care and attention. This is when you will begin to "see" the layers and begin to process information about the snow.

FIGURE S4-1. Having fun in a snow pit. At the left is the end of a snow coring tube. This snow scientist has a 100-cc density cutter in his right hand, his field book in his left, and the shovel and snow pit kit behind him. The ruler against the face has been set with zero at the bottom.

(Photo by Matthew Sturm.)

4. Use a card or palette knife and slide down the pit face; feel where the layers are harder and softer.

5. Do this again (or even a third time), and see if the differences in hardness correspond to visible differences. Insert markers (wooden coffee stirrers, straws) at layer boundaries.

6. If the layers are elusive, consider gently whisking the face. Brush both up and down and sideways. Avoid brushing in stratigraphy. Differences in grain bonding result in differences in resistance to brushing and layer boundaries will appear almost by magic. Repeat steps 4 and 5 after the brushing.

7. Begin to make measurements. Place a vertical ruler along the pit face, zero at the bottom (unless you are on a glacier, where there is no bottom). Record where you are, the date, your companions, and pertinent weather information (if you are digging in a blizzard, you will know the data are perhaps of lesser quality).

8. Start with temperature first as it does the least damage to the snow. Using a digital or dial thermometer, work downward, placing the instrument and then recording the temperature and the height.

9. Continue with the hand hardness test (penetration of a. a fist; b. four fingers; c. one finger; d. a pencil; and e. a knife). This is a test where you push something into the snow to get a relative feel for whether a layer is strong or weak.

10. Continue with density measurements. I key these directly to layers and often make several in a layer; some observers do continuous profiles.

11. Measure grain type and size using a comparator card with a well-marked grid.

12. Take photos of grains and the pit wall, but first take photos of the field book or the data sheet to ensure the photos get tied to the right pit.

13. Continue with specialized measurements.

14. Sit back. Look at the now partially destroyed pit. Anything you forgot?

15. Backfill the pit.

A few additional recommendations:

1. Find and adopt a protocol and format that you like for recording what you see. There are dozens of standardized formats out there. *The International Classification for Seasonal Snow on the Ground* (https://unesdoc.unesco.org/ark:/48223/pf0000186462), mentioned earlier in this chapter, provides a wealth of information and guidance and is a good place to start. There are even computer-based recording and plotting programs like Snow-Pilot (https://snowpilot.org/) that can be used.

2. Before digging, decide if you will be coming back at a later time to dig another snow pit or trench. Leave room if you are, and be careful not to disturb snow that you may need to measure later in the winter.

3. Decide which direction you will orient the pit wall. Be careful not to step or pitch snow where the top of the measurement wall will be. That surface is your reference line. Consider the sun angle now and where it will be in two hours when you are finishing the measurements and adjust the angle accordingly.

5

METAMORPHISM: SNOW IN THE CRUCIBLE

We think of snow as a "cold" material, and it is cold when compared to the temperature of the human body or a boiling kettle. But another way to think of snow is that it is never at a temperature much lower than its melting point of 0°C (32°F). The lowest temperature ever recorded on Earth was -89°C (-129°F) at Vostok Station, Antarctica. That is brutally cold, but the snow there was still less than 90°C below its melting point. A cold day in the Arctic is about -40°C (-40°F) with the snow just 40°C (72°F) colder than the temperature at which it melts. Iron melts at over 1,500°C (2,732°F), so at room temperature iron is over 1,470°C cooler than its melting point. Heat the iron up until it is 40°C below its melting point and it will glow yellow and bend and deform as easily as rubber (Fig. 5-1). Brought in contact with other pieces of hot iron, it will fuse with them. When it is yellow-hot, iron is just as volatile as snow. This analogy helps explain why snow grains are so good at shape-shifting: everywhere on Earth, they are close to their melting point.

FIGURE 5-1. A red-hot iron bar behaves differently than cold iron because it is close to its melting temperature. Snow, which is always near its melting point, behaves in a similar way, which is why it metamorphoses and changes so readily.

(http://farm1.static.flickr.com/22/26781770_ff25074252.jpg?v=0.)

5.1 METAMORPHIC AGENTS

Four main agents drive snow metamorphism (Fig. 5-2):

1. **GRAVITY:** The weight of the snow itself can produce forces that exceed the breaking point of fragile snow grains and which help to pack grains closer together. This results in snow settlement and densification.

2. **TEMPERATURE GRADIENTS:** Differences in temperature from one part of the snowpack to another (gradients) drive water vapor from one grain to another through sublimation and condensation processes. Gradients can drive water

vapor from one layer to another as well, altering the density of both layers. Gradients are symbolized here by juxtaposing a campfire (hot) and icicles (cold).

3. **WIND:** The wind can blow the snow about, pulverizing grains, yet ultimately make some of the strongest layers of snow (called wind slabs) through a process called sintering.

4. **HEAT AND WATER:** The sun and above-freezing temperatures can heat snow grains and melt them, producing water that infuses into the pack. Rain-on-snow can do the same. The water creates rounded grains, ice lenses, and icicle-like features embedded in the snow called percolation columns.

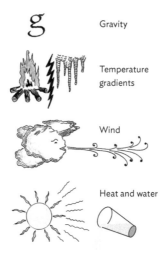

Gravity

Temperature gradients

Wind

Heat and water

FIGURE 5-2. Four agents that alter snow on the ground and drive snow metamorphism.

The four agents operate on an enormous number of grains at all times. There are about 500,000,000 snow grains per cubic meter (about 14 million grains in a cubic foot) for typical snow. A football field covered by 50 cm of snow (20" deep) would contain over a trillion grains (10^{12} grains). No two grains in the snowpack undergo exactly the same metamorphism, but statistically there are distinctive metamorphic pathways (section 5.2) that individual snow layers and grains follow. These ensure that layers will have characteristics that distinguish one layer from another even after metamorphism has occurred.

5.1.1 Gravity

Once they are on the ground, snowflakes do not last long in pristine form. Upon impact, or within hours of landing,

FIGURE 5-3. This eclectic collection of newly fallen snowflakes includes stellar dendrites, heavily rimed dendrites, and one hexagonal plate (center of picture). Three of the dendrites have already lost an arm (yellow arrows), broken off as they landed. (Photo by Matthew Sturm)

they begin to break apart under the pull of gravity. Stellar dendrites, with their long, thin arms, are particularly prone to breakage (Fig. 5-3). Even after the arms have broken off, gravity continues to try to compact the grains, pushing them closer together toward the cubic packing arrangement shown in Table 4-2.

5.1.2 Temperature Gradients

Gradients exist everywhere in nature. The wind blows because there are high- and low-pressure cells creating a pressure gradient. Red dye dropped into water diffuses because there is a chemical concentration gradient. Heat is transferred when two bodies with different temperatures are brought into contact because of the temperature gradient.

We are all familiar with temperature gradients, but not perhaps how they are calculated. Think of a house in winter (Fig. 5-4). The furnace keeps the inside at 20°C (68°F) even when it is -40°C (-40°F) outside. Across the walls of the house there is a 60°C (108°F) temperature drop. If the walls are 20 cm (8 inches) thick, then a temperature gradient of 3°C per centimeter (about 14°F per inch of wall) exists across the wall. It is this gradient that drives the heat from the house toward the outdoors, and drives up the heating bill. One way to reduce that bill is to make the walls thicker, which will reduce the gradient and diminish the heat flow. Another way is to use better insulation in the wall.

Virtually all snowpacks, except those that are wet (which are isothermal), are subject to temperature gradients. The gradients arise because the ground is warm from summer heating and wetting, while the winter air is cold. It is not uncommon to find the temperature at the base of the

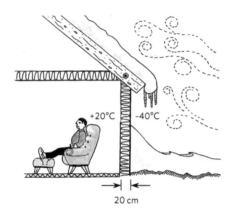

FIGURE 5-4. The temperature gradient across the walls of this cozy house in winter is -3°C/cm, computed by dividing the 60°C temperature drop (108°F) by the 20 cm (8") wall thickness. The gradient drives heat from inside the house to the cold outdoors. The thicker the wall, the lower the gradient and the less heat lost.

snow 10° to 20°C (18 to 36°F) warmer than the top of the snow, with such conditions lasting for weeks at a time. Temperature gradients drive water vapor sublimation and condensation, producing rapid changes in the size and shape of snow grains, and redistributing mass from one snow layer to another. Snow scientists have found that when the magnitude of the temperature gradient is greater than about -0.2°C/cm (about -1°F/inch), an ornate style of metamorphism, called **kinetic growth**, takes place. Kinetic growth produces depth hoar and surface hoar crystals, which we will examine shortly. When the gradient magnitude is below the threshold, equilibrium growth takes place, producing smoother, more rounded grains.

5.1.3 Wind

Snow deposited on the ground will start to move when the wind gets up to about 5 meters per second (11 miles per hour). In principle, this is not a sufficiently strong wind for moving snow grains at rest on the surface, but a leapfrog-like process called ***saltation*** (Fig. 5-5a) allows the wind to get the grains moving at this lower speed. Saltation happens because the wind a few centimeters above the snow surface is faster than the wind right at the surface. It works like this: an odd grain of snow, either a bit lighter than its neighbors or sticking up a bit more, is caught by the wind and starts rolling and then bounding downwind. As this first grain moves, it strikes another grain. The momentum of the first grain is just enough to pop the second grain up above the snow surface, where it is now caught by the slightly faster wind, and that starts the second grain moving downwind. Sometimes the first grain might even pop two grains into the air at once, and when they fall back to the snow surface, they can eject four or more grains. Soon all of the surface snow is in motion. The jumping grains follow beautiful parabolic trajectories that a Japanese snow scientist, Dr. Daiji Kobayashi, filmed in the 1970s (Fig. 5-5b).

When the wind gets stronger, the snow grains leap higher and are blown farther. At about 10 meters per second (22 miles per hour), many saltating grains begin to break apart on impact, creating smaller and lighter grains. Instead of dropping, some of these grains are lofted by the wind, and as long as the wind continues to blow, they don't come down. They are so light, they float and swirl on the turbulent currents, in some cases rising more than 50 m (164 feet) above the ground. These high-flying grains are called the

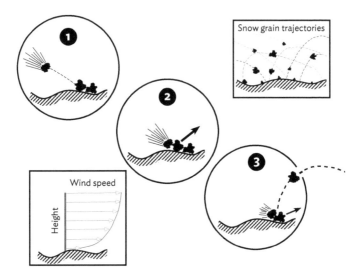

FIGURE 5-5a. (Numbers 1 through 3) Saltation is a process by which a snow (or sand) grain entrained in the wind ejects other grains upon impact, sending them up into the stronger wind stream that exists higher above the surface (inset lower left). There the grains are entrained in the wind and fly downwind, where they eject multiple other grains, until the entire layer of grains is bouncing and traveling downwind (inset, upper right).

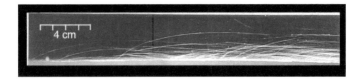

FIGURE 5-5b. Saltating snow grains photographed by D. Kobayashi in a wind tunnel in Japan following beautiful parabolic trajectories.

suspended load. A haboob (a desert dust storm) happens when the wind lofts quartz grains into suspension. Haboob particles are 0.01 to 0.05 mm in size. Ice particles are about one-third the mass of quartz, so the size of suspended snow grains can be larger, closer to 0.1 mm (4 thousandths of an inch) , because the force of gravity pulling them down is less. A combination of saltating and suspended snow grains will produce full-on blizzard and white-out conditions that can range from unpleasant to truly life-threatening (Figs. 5-6a and 5-6b).

During one of our Arctic expeditions, we got caught in a severe blizzard while traversing a long and exposed ridge. We had to get off the ridge and find shelter. Thankfully, we had a good map and a GPS, which allowed us to find and navigate our way off the ridge and down into a sheltered valley. We got the tents up, and finally we could relax. Looking across the valley, I couldn't see more than 50 m (about 164 feet), yet looking straight up, I could see the sun shining down on us, albeit looking a bit hazy. It was not snowing, yet our clothing and equipment were rapidly being covered in a powdery substance that looked like snow, but had the consistency of powdered sugar. It clung to surfaces in ways that normal snow never does. This material was the suspended snow load falling out of the sky, pulverized by the wind so that it was as fine (and as small) as diamond dust. The odd clingy behavior of the snow was due to its unusually small particle size, and perhaps a static charge.

5.1.4 Heat and Water

The snowpack can be heated directly by the sun, by a warm weather front with above-freezing temperatures (sensible

FIGURE 5-6a. (Left) Two snow scientists shelter behind a snowmobile during a blizzard in northern Alaska. Their clothes are covered by finely pulverized drift snow, a part of the suspended load that is falling out of the wind in the lee of the snowmobiles.

FIGURE 5-6b. (Below) Both saltating and suspended grains are moving in this picture taken in the Northwest Territories of Canada. The smoke-like streamer rising to the person's knees is the saltating load; the overall haze tells us there is a suspended load as well.

(Photos by Matthew Sturm.)

heat), from the ground beneath the snow (ground heat), from rain falling on the snow, and from condensation of water vapor at the snow surface (latent heat). If the snow is already at 0°C, any additional heat introduced from these sources will cause melting and introduce liquid water into the snow. Scientists term this "excess heat."

When water moves through the pores of a snowpack, the process is complicated. The fate of the water is determined by how much water is flowing, the temperature of the various layers of snow (e.g., their cold content), and the hydraulic conductivity (K) of the layer, which is a measure of how easily water can move through the connected pores. Due to the cold content, some water freezes as the liquid flows through the pores, creating ice lenses, layers, and percolation columns. At the same time, the flowing water that doesn't freeze tends to round and coarsen the snow grains it flows past. This coarsening process increases K, thereby producing a positive feedback: the more water flowing, the more the snow grains coarsen. The more they coarsen, the more easily water can flow.

At some point, if enough water has been injected into the snowpack, it will have warmed the pack throughout to its melting temperature, 0°C. We call snow in this state "isothermal," and any additional introduction of water will begin to melt the bonds between grains. Now, instead of a cohesive pack, the snow becomes a slush, a mixture of icy grains floating in water. Slush seems to be universally hated. It makes for bad driving and wet shoes (Fig. 5-7a), but it also results in interesting elliptical particles (Fig. 5-7b) that continue to coarsen over time.

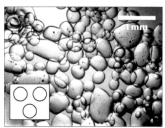

FIGURE 5-7a. When slush develops from snow, it means wet shoes and bad driving.

FIGURE 5-7b. When we look closely at what is in the slush puddles with a microscope, some wonderfully strange crystals are found. The symbol for this type of snow, three separated circles, appears in the lower left of the picture of the slush grains.

(iStock photo 902516106 (left). Micro-photo by Sam Colbeck (right). Used with permission.)

5.2   METAMORPHIC PATHWAYS

The agents described above produce four distinct metamorphic pathways of change (Fig. 5-8). Three of these pathways are "dry," meaning no liquid water is present, while the fourth involves liquid water.

5.2.1 Wet Snow Metamorphism

Water, whether due to melt or rain, generally enters the snow at the top of the pack and from there percolates downward. It does this in distinct flow fingers rather than seeping down in a broad wetting front. Scientists still lack a full understanding for why this type of flow instability occurs, but it is not unique to snow. Dye tracer and laboratory experiments confirm similar paths develop when water percolates into dry sand (Figs. 5-9a and 5-9b).

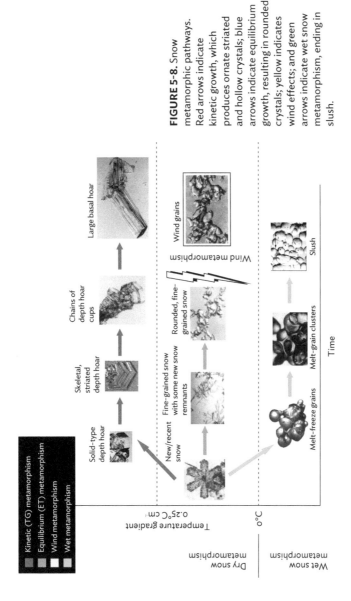

FIGURE 5-8. Snow metamorphic pathways. Red arrows indicate kinetic growth, which produces ornate striated and hollow crystals; blue arrows indicate equilibrium growth, resulting in rounded crystals; yellow indicates wind effects; and green arrows indicate wet snow metamorphism, ending in slush.

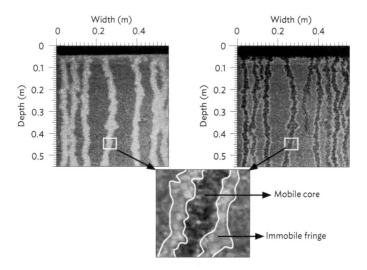

FIGURE 5-9a. Flow fingers percolating into sand have been made visible using dyed water. The fingers are a few centimeters wide and about 10 cm (4") apart, similar to what is observed in snow. In close-up, they have a central core where the water was flowing, and an immobile fringe where it permeated the sand and then stagnated. In snow that fringe freezes to produce hollow ice pipes with liquid water percolating down the center (see Fig. 5-9b). (Fig. 5-9a from F. Rezanezhad. Used with permission.)

In snow, though not sand, when the downward flowing fingers encounter changes in grain size (and changes in hydraulic conductivity, K) due to snow layering, or ice layers that were created in previous melt/rain events, the water spreads out and ponds within the pack. Ponded water will eventually find a weak spot (or drain) where it can break through and continue downward again, but not before some of the ponded water has refrozen, creating one or more icy layers or icy lenses within the snowpack.

FIGURE 5-9b. An icy percolation column excavated from a snow pack that had experienced a rain-on-snow event. The geometry is similar to the sand flow fingers shown in Figure 5-9a, but in addition there are remnant edges of ice lenses sticking out horizontally in several locations where water had spread out in the snow then froze. For scale, the whisk broom is 20 cm long.

(Photo by Matthew Sturm.)

Dye experiments are a good way to visualize water moving in snow, and they tend to reveal some aspects of the process that are not immediately intuitive. For example, Figure 5-10 shows how a single (artificial) wetting event produced multiple ice layers within the snowpack, not just at the top of the snow. Similarly, the event that created the percolation column in Figure 5-9b created three icy layers simultaneously. A spray bottle or a garden sprayer filled with water mixed with nontoxic dye, a shovel, and time are all that is needed to carry out this type of experiment. Dyed water is sprayed at the top of the snow and moves downward, creating percolation columns and ice lenses that can be readily seen in profile when a snow pit is dug adjacent to the sprayed area (see also Fig. 5-11). A word of caution, however: water likes to migrate to an open face in snow, where it tends to collect, giving the impression of far more water flow than is actually taking place. Shaving the vertical face back using the shovel can clarify the real size of the flow features. Even after conducting this experiment many times with students and school children, I still get surprised at how fast the water can move down through the snow and arrive at the base of a snowpack. Sometimes it takes just a couple of minutes, evidence of how efficient those internal flow fingers are at moving water.

In addition to making icy layers, water flowing through the snow causes the grains to round and coarsen (Fig. 5-12), as was discussed for slush. The grains first lose their sharp edges, then they begin to grow, the larger grains scavenging ice material from nearby smaller grains. Eventually, the grains become smooth ice spheres that cluster together (Fig. 5-13). With more melt-freeze cycles typical of spring

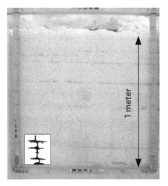

FIGURE. 5-10. A 1-by-1-meter (about 3 by 3 feet) block of snow into which dyed water has been introduced at the top. In seven places the water has ponded and spread out. All of these will end up being ice layers when the water refreezes, and none of them formed at the surface of the pack where the water started. The symbol denoting percolation columns appears to the right.
(Photo from M. Williams, used by permission.)

FIGURE 5-11. We rarely get to see the distribution of percolation columns in the snow after a melt event, but here the tops of hundreds of such columns are visible. The snow had been melting for a few days, producing water that had percolated down to the base. Then it got cold again and everything refroze. Next, the surface snow settled a bit, but it could not settle as readily where it was supported by the icy percolation columns, which now appear as bumps at the top of the snow. The columns were spaced 10 to 15 cm (5") apart.

days and nights, these spheres fuse even more fully, eventually becoming even larger polycrystalline masses like those shown in Figure 5-14. Skiers know this type of snow as **corn snow**. In spring, when the days are well above freezing but the nights are still cold, the icy bonds holding the rounded grains melt as the day progresses. By late afternoon, the snow is a pleasant mixture of rounded ice grains lubricated by generous amounts of meltwater, not bad to ski on.

FIGURE 5-12. These water-coarsened grains that are over 1 cm (0.4") in diameter were the result of several weeks of daily snowmelt and water percolation, followed by refreezing at night.

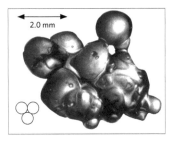

FIGURE 5-13. These clustered rounded grounds are known as melt clusters. Note that there are thick necks or bonds between the grains, but the grains are still identifiable as individual particles. The symbol for this type of snow appears at the lower left.

(Sam Colbeck. Used with permission.)

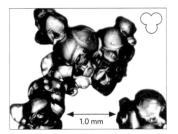

FIGURE 5-14. Rounded polycrystals. Here the grains have fused considerably more than in Figure 5-13. There are no longer distinct necks connecting the grains because simply too much water has been introduced. The symbol for this type of snow appears at the upper right.

(Sam Colbeck. Used with permission.)

Water moving through the snowpack also acts at scales larger than grains and layers. A common sight in the mountains in spring is to see slopes looking like Figure 5-15, where rills have developed within the snow. These drainage features run perpendicular to the local contours and represent internal downslope drainage channels. In the rills (the depressions between ridges) the grains have coarsened, increasing K, so the water tends to flow in these depressions rather than migrating into or across small snow ridges.

FIGURE 5-15. Water rills in a spring snowpack, channeling water downslope. (Public domain.)

Sometimes just a little melting will take place right at the surface. Perhaps it will be warm enough for mist or even rain. The snow surface will become hard and shiny. In Iñupiaq this is called:

silligruaq

The Germans have a nice word for this:

firn-spiegel.

It means ice mirror.

FIGURE S5-1. Firn-spiegel.

FIRN-SPIEGEL

In April 2007, we were traveling east across the Barren Lands of Canada. As we moved east, we came to an area where a week before our arrival there had been freezing rain. This had produced an ice crust on the top snowpack. The rainstorm had evolved into a snowstorm toward the end, so the icy surface, the firn-spiegel, was covered when we first encountered it. But after a few miles, we came to an area where the wind had swept the new snow from the higher icy bumps (Fig. S5-2). A little farther on, even more of the icy surface was exposed. Walking became treacherous because the icy layer was strong enough to support our weight, but was so slick it was almost impossible to get any traction on it. We all fell several times before we learned to step only in places where the firn-spiegel was covered with the new snow. As the Earth's climate warms, there is some indication that adverse conditions like rain-on-snow may increase in frequency.

FIGURE S5-2. Firn-spiegel peeking through a covering of new snow in Northwest Territories, Canada. (Photo by Matthew Sturm.)

5.2.2 Wind Metamorphism

During wind-driven saltation, snow grains get beat up. They break and chip through impact, creating smaller compact grains with sharp edges and irregular points (Fig. 5-16). These grain remnants can pack together tightly, and where they touch, they form strong bonds. It is this bonding that produces wind slabs, those cohesive snow layers that give wind-drifted snowpacks their essential character and which can be cut into blocks and made into igloos.

So how does a cornice made of wind-driven snow curl like a wave without falling apart or breaking? The answer is a process called *sintering*. It is a process unfamiliar to most people, yet many of the parts in a cell phone are produced by sintering. Pulverized metal is packed into a mold and heated,

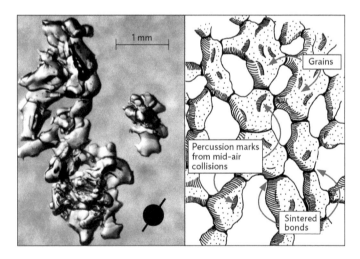

FIGURE 5-16. (Left) A cluster of wind grains. Note the fracture surfaces and thick bonds. The symbol for this type of snow is shown in the lower right corner of the left image. (Right) Wind slabs are strong because the bonds are thick and numerous, creating a well-sintered network structure.

but not hot enough to melt it, then the hot metal dust is allowed to cool. When removed from the mold, the metal will have bonded into a solid mass. This happens because metal molecules migrate to everywhere that the particles are touching, and there the molecules are deposited. This molecular deposition results in thick bonds. Snow sinters in the same way as metal, except unlike metal sintering, heating the snow is unnecessary because it is already close to its melting point (Fig. 5-1). When wind-pummeled snow grains sinter, the bond necks that form can be almost as thick as the grains themselves, resulting in a very strong material.

Snow sintering happens in hours to days. Immediately after the wind stops blowing, the drifted snow will be soft and easy to shovel, soft enough for a boot to sink into, but twenty-four hours later, the same snow can be cut into solid blocks with a snow knife or saw (Figs. 5-17a and 5-17b). Once drifts have hardened like this, the snow can no longer be moved by the wind, even when it starts to blow again. One of the school demonstrations we do related to sintering is to suck snow into a cold shop vacuum. This does a fine job of pulverizing the snow. We then pour the snow from the vacuum canister into molds. Three to four hours later, *voilà!* The snow is a solid mass that has taken on the form of the mold. A quinzhee shelter is possible for the same reason: by piling up snow with a shovel, the grains are pulverized and therefore sinter well, allowing the mound to be hollowed out after a little time has elapsed.

The combination of wind-driven snow transport and sintering results in a wide variety of drift features (Fig. 5-18) including:

FIGURE 5-17a.
A massive snow
cornice in the Alps.
Without sintering
making strong bonds,
snow could not
overhang like this.
(Public domain.)

FIGURE 5-17b. This
snow wall has been
built to shield the
orange tent from the
wind. The blocks of
drift snow have been
cut with a snow knife
(seen just inside the
wall). Sintering has
made the blocks quite
strong and cohesive,
and has also provided
the "glue" that allows
the blocks to lean in
without falling over.

- Snow waves
- Dunes, barchans, and whalebacks
- Ripple marks
- Pits and pockmarks
- Snow steps
- Sastrugi
- Cornices
- Ribbon drifts

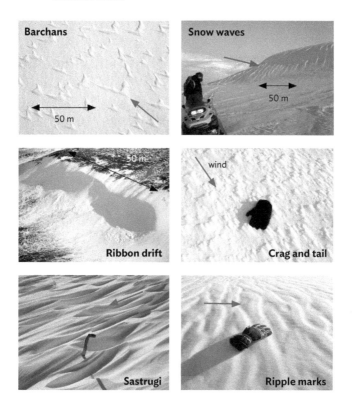

FIGURE 5-18. Some of the many large and small drift features produced by the wind. The red arrow shows the direction of the wind.

(Photos by Matthew Sturm and Simon Filhol. Used with permission.)

FIGURE 5-19. Endless fields of sastrugi in Nunavut, Canada. The wind that cut and eroded the snow originated from behind the photographer's back. (Photo by Matthew Sturm.)

Much of my snow experience is in the Arctic, where wind-drift features are everywhere. While snow dunes, ribbon drifts, and the other features in Figure 5-18 are common, the most widespread feature, and the bane for us when traveling, is sastrugi (Fig. 5-19). These features are erosional forms, the cut-up remnants of former dunes, whalebacks (dunes that look exactly like the name implies), and barchans (Fig. 5-18). They can range from moderately hard to rock-hard, and a few hours of driving over hard sastrugi can make it feel like your teeth are going to fall out.

Snow cornices (Fig. 5-17a) and ribbon drifts (Fig. 5-18, left middle) are created by a wind rotor, a horizontal vortex in the wind that forms on the lee side of a bluff or downwind from the edge of a cutbank (Fig. 5-20). Saltating snow grains get blown along the snow surface until they are swept over the edge, where the rotor wind curls the grains down and under. Sintering "glues" the grains together, and soon the snow has mimicked the shape of the rotor wind itself. With time, the weight of the cornice will cause it to droop and settle, accentuating the wave shape even more. In fact, it is not unusual for a cornice crack to form a few feet back from the edge due to the enormous weight of the snow. Cornice cracks have the potential to be dangerous, as they can be hidden by a thin layer of snow and unwary travelers can fall into the crack and get stuck.

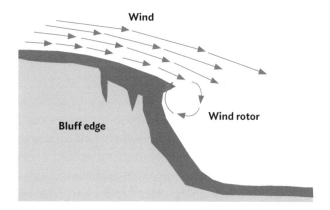

FIGURE 5-20. When the wind blows over the edge of a gully, it forms a wind rotor (a horizontal vortex) because the wind cannot bend steeply enough to follow the surface of the land. Saltating snow grains driven by the wind get swept over the edge and into the rotor, which creates essentially a wave of snow, better known as a **cornice**. Once formed, the snow cornice sags or bends downward under its own weight, accentuating the rotor shape.

5.2.3 Temperature Gradient Metamorphism: Sublimation and Condensation

Even when a snow cover remains below freezing (no water) and isn't being blown about, it will still undergo metamorphism due to temperature gradients. This is called ***dry snow*** metamorphism. Dry snow metamorphism at first glance seems more mysterious than the other pathways because it takes place through water vapor transfers driven by temperature gradients, and it happens one water molecule at a time. We cannot see these processes in action, hence the mystery. Yet everyone has seen evaporation at work, and that too is the result of vapor pressure gradients. Set a pan of water out on a hot day and all of it can disappear in twenty-four hours, the water going off one molecule at a time. The warmer the water, the faster the evaporation rate. The same happens for ice, though the transfer rate is slower, and the name of the process is now ***sublimation*** because the molecules are going from a solid (ice) to a gas (water vapor).

To understand how sublimation leads to snow metamorphism, and how it is connected to a temperature gradient, it is useful to imagine that you are tiny, smaller than a grain of snow, and that you are crawling through a matrix of snow grains in the snowpack, through the linked series of pore spaces between grains. Your world might look something like Figures 5-21a and 5-21b.

The ground beneath the snow is warmer than the air above, so heat is flowing up from the bottom to the top of the pack. The grains and crystals are made of ice, and these conduct heat nicely, but the pores are filled with air, which doesn't. Standing in a pore space, with your feet on one grain and your hands touching the grain above your head, you

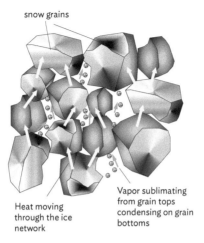

FIGURE 5-21a. A schematic showing ice grains and air spaces. The yellow arrows indicate the conductive heat pathways through ice grains. The small spheres indicate water vapor sublimating from ice grains and migrating upward through pores.

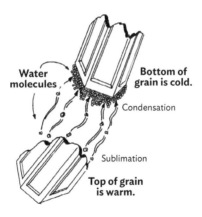

FIGURE 5-21b. The water vapor sublimates from the top of the lower grain and condenses on the bottom edge of the higher grain, rounding the top edge of the lower grain, while creating a razor-sharp lower edge on the upper grain.

would notice a distinct difference in their temperatures. The upper grain would be colder. Water molecules sublimating from the lower, warmer grain would travel up across the pore space and condense when they came in contact with that colder surface. The bottom edges of the upper grains would grow, while the top edges of the lower grains would shrink. And indeed, this is exactly what we see happening in real snow (Fig. 5-22). Another Japanese snow scientist, Dr. Zyungo Yoshida, had a charming name for this process: "hand-to-hand" growth.

There is a surprisingly abrupt split in the dry snow metamorphic pathways and the grains they produce (see Fig. 5-8, red versus blue arrows), based on the rate at which sublimation and condensation are taking place. That rate is controlled by the temperature gradient across the snow: when that gradient is high, the flux of water molecules to growing grains and crystals is so rapid, ***kinetic growth*** takes

 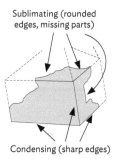

Sublimating (rounded edges, missing parts)

Condensing (sharp edges)

FIGURE 5-22. A snow crystal from the base of the Fairbanks snowpack. The bottom edges and front corner are razor-sharp, indicating they have been growing. The top edges are rounded, eroded, and large sections of what was once a hexagonal cup crystal are gone, indicating considerable sublimation losses. (Matthew Sturm.)

place (e.g., crystallization processes control the growth rate). The result is ornate hexagonal crystals with striated faces. Broadly, these are called **faceted crystals** (they have planar surfaces and 120° angles) or **depth hoar**. But if the gradient drops below the critical value of about 0.2°C/cm, then kinetic growth stops and is replaced by **equilibrium growth**. This pathway produces smooth, rounded grains reminiscent of those produced when water is present, but formed without any liquid water being present.

The existence of equilibrium growth was first demonstrated in a classic snow experiment. The scientists trapped a single stellar dendrite in a glass jewel case and sealed it. Then they held that crystal at a fixed temperature, careful to be sure there were no temperature gradients across the case. After fifty-seven days, the same amount of ice was present as before, but now there was a series of rounded equilibrium ice grains (Fig. 5-23) rather than a faceted, branched crystal.

The transition between the kinetic growth of highly faceted, striated crystals with hexagonal symmetry, and the equilibrium growth of rounded grains arises out of two related processes: vapor supply and crystal growth kinetics. Under a strong gradient, there is so much water vapor available, far more than could ever be incorporated into a growing crystal lattice, that the orderly packing of molecules into the lattice dominates the rate and style of growth. That is why the crystals have recognizable hexagonal forms. During equilibrium growth, on the other hand, a growing crystal is starved for vapor, so the Kelvin effect (Chapter 2) and the minimization of surface free energy dominate the growth and a smooth, rounded crystal is the end product.

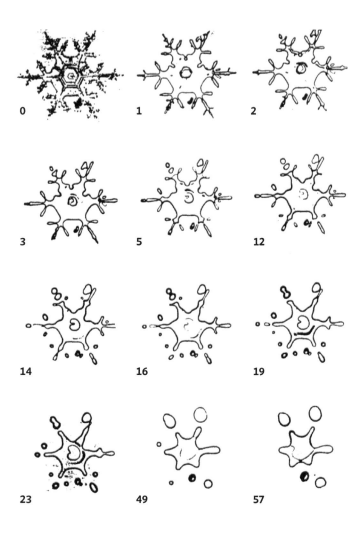

FIGURE 5-23. A stellar dendrite held at a constant below-freezing temperature with no temperature gradient slowly rounds and breaks into smooth, round ice grains (equilibrium growth). The numbers are the days since the start of the experiment.

(From H. Bader, "Snow and Its Metamorphism," SIPRE Translation No. 14. U.S. Government, 1954.)

Equilibrium grains (Figs. 5-24a and 5-24b) are probably the most common type of snow grain when considering all snowpacks, yet they seem to be the least photographed and noted. That is not surprising: compared to their exotic cousins (faceted grains, depth hoar, melt clusters, polycrystals) equilibrium grains are plain. But they are worth noting since so much of the snow we see and ski on is comprised of this material.

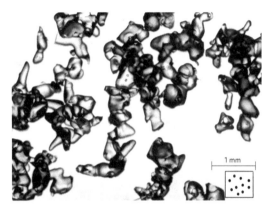

FIGURE 5-24a. Classic rounded equilibrium growth grains. The symbol for this type of snow (lower right) is a series of dots; smaller dots suggest finer grains. (Micro-photo by Sam Colbeck. Used with permission.)

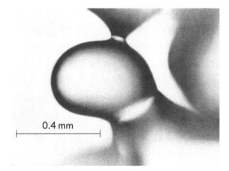

FIGURE 5-24b. The end state of equilibrium growth is a smooth, highly rounded grain. (Micro-photo by Sam Colbeck. Used with permission.)

Kinetic growth produces the show-off grains shown in Figure 5-25. These can have simple smooth facets, or they can be striated, hollow and cupped, even scrolled. Collectively, these types of crystals have been called *depth hoar* because kinetic growth produces its most extreme changes at the base of the snowpack where the snow has been in place the longest and is also the warmest. But depth hoar can also be found higher in the pack when conditions are right. The word *hoar* indicates a large, feathery type of crystal, and there is a regular progression of crystal forms from solid crystals with incipient facets that look like cut glass, to hollow bell-like cups, to vertical chains of cups. Faceting by kinetic growth typically robs bonds to grow the crystal faces, so depth hoar layers are notoriously brittle and weak. They often serve as the failure plane for an avalanche.

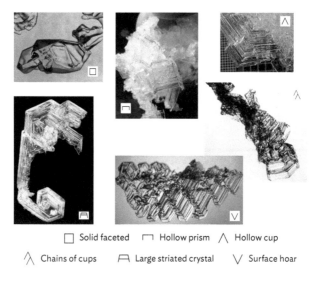

☐ Solid faceted   ⊓ Hollow prism   ∧ Hollow cup

∧ Chains of cups   ⊐ Large striated crystal   ∨ Surface hoar

FIGURE 5-25. Showy snow grains resulting from kinetic growth. The symbols for each type are shown. (Photos by Matthew Sturm.)

5.3 SNOW LAYERS AND GRAIN TYPES

5.3.1 Climate Classes and Metamorphic Pathways

During their existence, real snow layers rarely track along a single metamorphic pathway as in Figure 5-8. When these layers form in early season, they may land on warm ground and undergo some basal melting (wet snow metamorphism), then the wind may come up and erode or move the grains (wind metamorphism). Later, as the nights get cold, the temperature gradient across the snow might go critical and kinetic growth (dry snow metamorphism) begins. Perhaps later in the winter, rain water will percolate down and wet parts of the layer. In essence, each layer tracks a path through Figure 5-8, but not always a simple path. This concept is analogous to the way a falling snowflake tracks a complex path through a Nakaya diagram (Fig. 3-8). Occasionally, though, nature does produce a snowpack that is the product of a single, simple pathway, or perhaps two pathways (Fig. 5-26). More often, a layer will have crystals and grains that tell a story of multiple pathways followed over the course of the winter.

The existence of snow climate classes (Figs. 1-4 and 1-5) tells us that there are associated groups of related pathways that can be observed more readily in one geographic location than another. Out on the Arctic tundra, we might not see a set of layers as simple as Figure 5-26, but we could expect to find mostly depth hoar and wind slab because the tundra is invariably a windy place with a thin snowpack across which strong temperature gradients prevail. In the Coast Mountains of British Columbia, we might expect a deep pack, and with air temperatures that hover around freezing and frequent rain-on-snow events, lots of wet snow metamorphism features in that pack (Fig. 5-27).

FIGURE 5-26. This typical tundra snowpack is about as simple stratigraphically as snow can be. A 15-cm-thick (6") snowfall landed on cold tundra, then was subjected to extreme temperature gradients for months, producing large, faceted depth hoar grains. A substantial second snowfall that has capped the first layer arrived accompanied by strong winds, producing a dense, well-sintered wind slab about 6 cm thick (2-1/2") that caps the depth hoar. (Photo by Matthew Sturm.)

FIGURE 5-27. A deep, wet maritime snow cover with numerous ice layers and melt features. (Public domain.)

Learning to read the snow takes practice, but again, little in terms of money or tools. A shovel, a whisk broom, a folding ruler, and a good magnifying glass or low-power microscope are enough (Fig. 4-4b). Below, Figures 5-28a and 5-28b are the front and back of a card we prepared for students taking our snow school. If the images are copied and printed so the grid is a true 2-mm square, then the two sides glued back-to-back and laminated, the card can serve as a useful reminder of the grain types, their size, and their symbols for sketching in a field book. It can help in reading the snow.

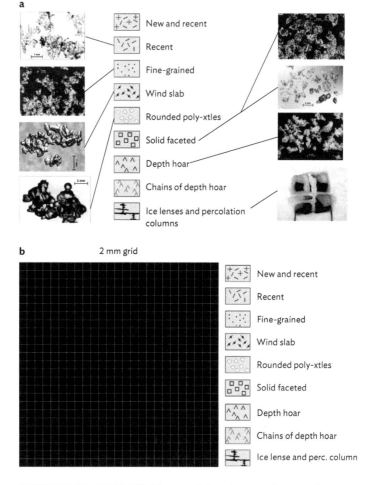

a

	New and recent
	Recent
	Fine-grained
	Wind slab
	Rounded poly-xtles
	Solid faceted
	Depth hoar
	Chains of depth hoar
	Ice lenses and percolation columns

b   2 mm grid

	New and recent
	Recent
	Fine-grained
	Wind slab
	Rounded poly-xtles
	Solid faceted
	Depth hoar
	Chains of depth hoar
	Ice lense and perc. column

FIGURES 5-28a AND 5-28b. These cards have been used at several Internation Snow Field Methods Schools and students found them useful. Scan or photograph the cards, mount back-to-back, laminate, and take into the snow. Be sure the 2 mm grid is actually a true 2 by 2 mm after the reproduction process.

6

THE GREAT WHITE BLANKET

Water molecules **to** *snowflakes, snowflakes* **to** *layers, layers* **to** *snowpacks, snowpacks covering vast fields, forests, mountains, glaciers, and frozen oceans:* the power of snow to affect climate and living things, including all of us, comes from both its microscopic characteristics and the vast extent of the Earth over which it spreads each winter (Chapter 1). In this chapter, we look at some of the more important properties of that vast white blanket.

For five years I chaired an international working group of snow scientists interested in improving our ability to monitor snow conditions and predict its impact on society. Within the group were two factions, one primarily interested in snow and climate, the other in snow and water. Both aspects of snow are crucial, but at times, let's just say these two groups had difficulties appreciating the other group's perspective. In an effort to broker peace, I spoke with a wise and very senior scientist about the issue, a man known for his pithy way with words. He thought for a moment, then said, "Tell them this: *Snow—it cools our planet and quenches our thirst.*" Neither a better nor shorter summary of the importance of the great white blanket in climate and hydrology is likely to be found.

Four of the important things snow does for our planet, for us, and for living things are:

- *Snow **reflects** sunlight.* New snow can reflect about 90% of the incoming solar radiation. Aged snow can reflect about 70%. Tundra, in contrast, only reflects 10%, forests about the same. Ocean water reflects even less, just 6% of incoming solar energy. Once a landscape is covered by snow, this superb reflective property ensures that the local air temperature will be lower because the snow will reflect sunlight back to space rather than allowing it to warm the surface. Snow reflectivity (known as **albedo**) plays a major role in keeping the Earth cool, and not surprisingly, is a key element in global warming.

- *Snow **insulates** the ground.* Because snow contains so much air (50% to 90% of a snowpack is air space) it makes a fine insulating blanket, nearly as good as a down parka or feather quilt. The ground beneath the snow is protected from frigid winter temperatures, as are any creatures that spend the winter under the snow. It is well known in cold climate regions that if snow doesn't arrive before the first cold snap hits, buried water pipes and septic systems will freeze. Houses that hold snow on their roofs are warmer than those that do not. Without a good snow cover, the ground would freeze harder and deeper in winter, and in the far North permafrost would be colder and thicker.

- *Snowmelt produces **runoff**.* It is estimated that over one billion humans rely on the water that comes from melting snowpacks for drinking, irri-

gation, industrial use, power generation, and recreation. Snowpacks in the mountain chains of the
world have been called the people's water towers,
and these towers currently cost nothing to build
or maintain. Fiscal estimates of the value of snowmelt runoff top trillions of dollars, and snowmelt
water provides additional value by maintaining
healthy ecosystems, including recharging wet-
lands—home to millions of birds, animals, and
fishes.

- *The mechanical properties of snow affect the way we
  function and **travel**.* Snow is slippery but cohesive.
  The slipperiness has led to the development of
  many innovative ways to travel over the snow, so
  humans living in snowy regions often switch their
  mode of travel in winter, using snowshoes, skis,
  snowmobiles, and dogsleds. Even in more tem-
  perate climates, people still change their tires to
  ones designed for driving on snow. Where changing our mode of travel isn't possible or acceptable,
  we try to get rid of the snow. The combination of
  snow depth and snow cohesion produces a material that is hard to drive through. Consequently, in
  order for wheeled vehicles to continue to be used,
  snow is plowed off roads, runways, and railways,
  an activity that costs billions of dollars each year.
  And the adverse effects of snow on transportation, which cause a significant number of deaths
  and injuries each winter, are tolerated because we
  have little other choice.

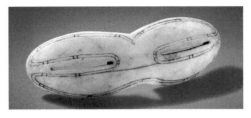

FIGURE 6-1. (Top) Carved ivory snow googles from Nunavut, Canada, circa AD 1200. (Bottom) Modern sunglasses. (Eskimo goggles: https://www.historymuseum.ca/cmc/exhibitions/tresors/ethno/et0174be.html. Modern sunglasses: public domain.)

6.1 REFLECTION

Stylish carved snow goggles (Fig. 6-1) have been found in Greenland and Siberia that are more than a thousand years old, and today we continue to be infatuated with sunglasses. These facts testify that humans have known for thousands of years that snow reflects light well, and that we need to protect our eyes when out in the snow. But a full understanding of snow and sunlight requires a deeper dive because snow both reflects and absorbs light in complex ways.

Figure 6-2 shows the electromagnetic spectrum with the visible light spectrum (red-orange-yellow-green-blue, etc.) in color. We are mainly interested in the visible and infrared sections when it comes to snow reflection, but is worth noting that scientists have tried using virtually every wavelength to

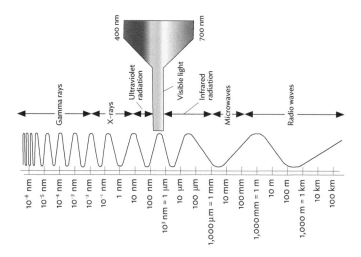

FIGURE 6-2. The electromagnetic spectrum with the visible spectrum shown in color. The spectrum is divided into sections based on the wavelength of the radiation. The abbreviation **nm** indicates nanometers (10^{-9} meters); **μm** indicates micrometers (10^{-6} meters).

map snow from aircraft or satellites. For example, gamma rays, which are produced by radioactive decay in most soils, are attenuated (or reduced) when the soil is covered by a layer of snow. By measuring the gamma count in autumn before it snows, and then later in winter when there is snow present, it is possible to determine how much snow water equivalent (SWE) is on the ground based on the reduction in the gamma count. The caveat that makes this method less useful for remotely sensing SWE is that gamma detectors that are sensitive enough to measure SWE rapidly have to be quite large, are heavy, and are quite expensive. They are therefore difficult to put on an aircraft or into space. Currently, most efforts to determine snow properties (like depth or SWE) from space are focused on the microwave range, because electromagnetic waves at those wavelengths excite water molecules and are therefore sensitive to how many of them there are. That is also why a household microwave can boil water.

Zeroing in on the ultraviolet (UV), visible, and infrared parts of the spectrum (Fig. 6-3), several features are important. The first is that peak incoming power of a solar ray (measured in watts/square meter) falls around 500 nm, right where the human eye is most sensitive (a typical eye responds to wavelengths from 380 to 740 nm), so sunglasses *are* a good idea. The second thing is that the Earth's atmosphere, which is filled with water vapor that can absorb solar energy, makes significant inroads in the amount of radiation that reaches the surface of the Earth. Atmospheric water vapor accounts for four distinct absorption bands (gaps) in the sea level light spectra. Ozone (O_2) and carbon dioxide (CO_2) also absorb light energy in the atmosphere before it gets to the surface of the Earth, the latter being the greenhouse effect.

This incoming sunlight would warm the Earth if it were all absorbed, so what we need to know is how much is absorbed versus sent back into space by the snow. That quantity is called the **albedo**, a term which today means the percentage of incoming light reflected by a body, but which comes from an ancient Latin word meaning "whiteness," suggesting perhaps that even the ancients were thinking about snow. Albedo is computed by first measuring the reflected solar radiation, next measuring the incoming solar radiation, then computing the ratio of the two (Figs. 6-4a and 6-4b). Alternately, two radiometers—one facing up, the other down—can be recorded simultaneously. The ratio, multiplied by 100, is the albedo in percent.

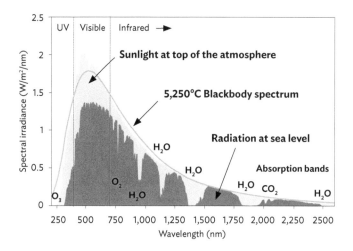

FIGURE 6-3. The UV, visible, and infrared spectrum showing the amount of light energy received at the top of the atmosphere (yellow) and at the surface of the Earth (red). The deep gaps are the wavelengths where water vapor (or CO_2 and ozone) in the atmosphere absorbs solar energy. The peak energy flux is in the visible wavelengths, 380 to 740 nm, the same wavelengths that human eyes can see. The temperature (5,250°C or 9,482 °F) is the surface temperature of the sun, and the gray curve is the theoretical output from the sun at the top of our atmosphere.
(https://en.wikipedia.org/wiki/Sunlight.)

Snow has one of the highest albedos of any natural surface on or above the Earth, its only serious rival being clouds (Fig. 6-5). When the snow is new and fresh, it can reflect nearly 90% of incoming light. When it melts away, exposing dark soil, forest, sand, or grass, the ability of the surface to reflect solar energy drops by a factor of 5 to 15X. If we revisit the snow cover maps in Chapter 1, they show that at times snow covers 50 million square kilometers of the Northern Hemisphere alone. When covered with snow, this area reflects an enormous amount of solar energy. That cooling

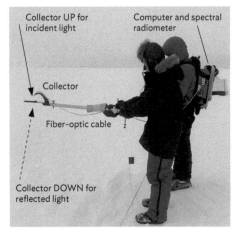

FIGURE 6-4a. (Left) Two snow scientists use a device called a field portable spectrometer to measure the spectral albedo of the snow (the albedo as a function of wavelength). The scientist will rotate the rod she is holding 180° after obtaining the value for incident light (light directly from the sun) so that she can then get the value for the reflected light from the snow. Using these two measurements, she will compute the ratio, which is called the albedo, a measure of the reflectivity. On an overcast day like the one shown here, the light is all diffuse (no shadows, no direct beams), and if the wind kicked up just a bit and snow started blowing, whiteout conditions would prevail. But despite the overcast, the snow is still reflecting over 80% of the incoming light.

FIGURE 6-4b. (Right) This snow scientist is carrying an albedometer that measures the incident and reflected light simultaneously (two radiometers, one facing up and one down). His hip boots and the slush around his legs suggest that snowmelt, driven by solar radiation, is well underway.

(Photo by Matthew Sturm [left]. Photo by Betsy Sturm [right].)

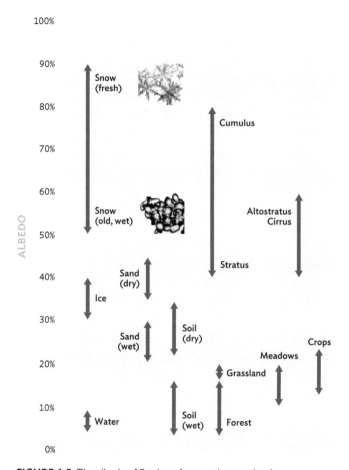

FIGURE 6-5. The albedo of Earth surfaces and some clouds.

effect is called *cryospheric radiative forcing*, and for the Northern Hemisphere it ranges from -2 to -5 watts/m^2 (-0.6 to 1.6 BTUs/hour/foot^2.).[1] The negative signs denote these are cooling rather than heating values, which for a warming planet, is a good thing. The troubling aspect of this remarkable cooling effect, however, is that reductions in the extent and duration of global snow cover on land and sea ice over the past few decades have reduced the effect by an amount estimated to be 0.5 watts/m^2 (0.16 BTU/hour/sq. foot). That sounds like a pretty dim light bulb's worth of energy, until we multiple that amount by millions of square kilometers where there used to be snow, but now there is not. Then the loss of snow equates with an increase in the solar heating rate of the Earth by 25 million megawatts, which is a whole lot of energy.

These calculations make it easy to see how crucial snow is to the cooling of our planet, and the nice thing about snow is that we don't have to do anything to achieve the effect. By way of comparison, in 1969 the installation artists Christo and Jeanne-Claude oversaw the wrapping of just 0.1 km^2 of the Australian coast with highly reflective cloth. That project took a month, one hundred workers, and 17,000 man hours. Nature wraps 500 million times that amount every year with a far more reflective material, and does it for free: we should appreciate this cooling effect, which our warming planet really needs.

Both aging (dry snow metamorphism) and melting (wet snow metamorphism) reduce the albedo of snow, with a mixture of rounded melt grains and liquid water producing an albedo value about half that of pristine new snow. Contaminants in the snow like black carbon also lower the snow albedo significantly.

[1] These figures are derived from a paper by Flanner et al. entitled "Radiative Forcing and Albedo Feedback from the Northern Hemisphere Cryosphere between 1979 and 2008," in *Nature Geoscience* 4 (2011): 151–55.

Ahh! If only it were that easy to understand snow and sunlight! In fact, to fully understand the role of snow and its albedo in cooling the Earth, we have to look at the albedo as a function of the solar wavelength, particularly in the infrared (thermal) region of the spectrum (700 nm and higher). At those higher wavelengths, the albedo runs from nearly 0 to about 0.5 (or 50%), meaning that instead of snow being a good reflector, it is actually a good absorber of energy at those wavelengths. As the snow melts away, exposing the bare ground, the reflection of infrared solar radiation actually goes up, not down (Fig. 6-6). At these wavelengths, also known as **longwave radiation**, snow has an emissivity of almost 1, which means it gives back as much energy as it gets. It re-radiates energy back to space extremely well, putting out energy at a rate proportional to the 4th power of the snow temperature. On a clear, cold night, an average snowpack might pump out 90 W/m^2 to space. This accounts for two common observations regarding snow. First, the coldest place on a cold, clear night is the surface of the snow. It is likely to be 5°C (9°F) colder than the temperature at eyelevel, so a night ski can produce cold feet while noses stay reasonably warm. Second, that cold surface is very effective at condensing water vapor out of the air, which produces surface hoar (Fig. 6-7), those beautiful feathery crystals that the next morning reflect the sun and make the snow world glitter and dazzle. Moreover, that same surface chilling due to longwave re-radiation produces huge temperature gradients in the top few centimeters of the snowpack itself. Very rapid kinetic growth can take place overnight, producing a phenomenon called **near-surface faceting** that is often the culprit in releasing avalanches.

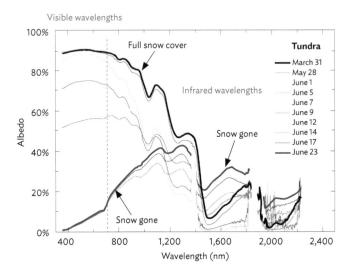

FIGURE 6-6. The evolution of the albedo of a snow cover as it melts in the spring. On March 31, the snow was 28 cm deep and covered all the tundra. By June 23, all the snow was gone. In the visible wavelengths (400–700 nm), the albedo dropped from 90% to 6%, but at 1,600 nm (infrared wavelength), it doubled from about 10% to 20%.

FIGURE 6-7. Surface hoar formation following a cold, clear night. (Public domain.)

6.2 INSULATION

Snow is generally very good insulation. Some types of snow, like new snow and depth hoar, are nearly as good insulation as commercial fiberglass batts. Other types are less so. For instance, wind slabs and dense icy layers are about six times more thermally conductive than new snow, which means they provide only one-sixth the insulation value. What this means is that under the same snowpack conditions, these latter types of snow would allow six times as much heat out of the ground. Snow scientists use a measure called the **thermal conductivity**, indicated by k, to describe the insulation value of the snow. It is the inverse of the well-known R-value marked on packages and sheets of building insulation. As k goes up, the insulation value (and the R-value) goes down.

Two factors influence whether a particular layer of snow will have low or high insulation value: the layer density, which controls whether the snow will contain more or fewer air pockets, and the grain bonding, which will determine how many solid ice connections exist between snow grains. These solid ice pathways (see Figs. 5-16 and 5-21) transfer heat far more efficiently than any other mechanism operating in the snowpack. New snow and depth hoar are both low density types of snow and are also known to be poorly bonded, while wind slabs and icy snow layers have thick bonds and fewer pore spaces. These structural differences explain the 4 to 6X differences in insulation value observed for various types of snow.

In Table 6-1 the k-values for snow can be compared to other common materials. Copper is a thousand times poorer insulator than wind slab, 4,000 times poorer than depth hoar. So in this respect, the table shows clearly that

FIGURE 6-8. The thermal blanket of snow is important in preventing hard ground freezing and protecting animals that live under the snow from frigid winter temperatures. This subnivean lemming is snug under his winter quilt of snow.

snow, all snow, is much more of a thermal insulator than a heat conductor (Fig. 6-8).

The last column of the table lists the thermal diffusivity. It is likely to be an unfamiliar term to many readers, but is a useful property when considering the impact of snow on the rapidity with which a cold snap could penetrate the ground and freeze a buried water line. Anyone who has made the mistake of holding the end of a copper rod while the far end was heated by a torch knows it has a very high thermal diffusivity: the temperature increases so fast in the unheated end, one can get burned. Copper is five hundred times better at propagating a temperature change than snow. In fact, the diffusivity value for snow is so low that for a 50-cm-deep snowpack (20") composed of both depth hoar and wind

TABLE 6-1. Thermal properties of snow and other common materials

Material	Density ρ (mks) [kg/m³]	Density ρ (cgs) [g/cm³]	Thermal conductivity k [W/m K]	Specific heat c [J/kg K]	Thermal diffusivity α [m²/s X 10⁻⁷]
Air	1.3	0.0013	0.02	1,005	153
Fiberglass batts	15	0.015	0.05	700	48
Depth hoar	250	0.25	0.10	2,108	2
Wood	450	0.45	0.15	1,700	2
Sheetrock (gypsum)	790	0.79	0.16	1,090	2
Wind slab	500	0.50	0.40	2,108	4
Water	1,000	1	0.58	4,187	1
Ice	917	0.917	2.30	2,108	12
Iron	7,850	7.85	55	449	156
Aluminum	2,712	2.712	205	897	843
Copper	8,940	8.94	401	385	1,165

Source: Matthew Sturm.

slab layers, when a cold snap hits, the bottom of the snow-pack does not feel the full temperature drop for more than a week. This is another reason the lemming likes to be down there at the bottom of the snowpack.

A practical example of where these thermal values have a big impact is the freezing of lake or sea ice. The ice freezes after the water has been cooled to 0°C, and it freezes at the rate at which heat can be withdrawn from the water. Early in the winter before it snows, the heat loss is through the skim ice that has formed (k~2.3), so the freezing continues rapidly. But once a layer of snow covers the ice, the escaping heat now has to move through the snow layer as well, with k about one-twentieth that of the skim ice, and the freezing process slows way down. The ice will continue to thicken, but now much more slowly. That is why snow is removed if people want thicker ice for skating on or harvesting for carving blocks. At a larger scale, this insulating property of snow plays an important role in the loss of Arctic sea ice. In early winter when it starts to get cold and the sea starts to freeze, if new snow falls on the ice, it will reduce how thick the ice will grow during the winter. With the Arctic ice pack forming later each year now, it is often quickly coated by snow, the result: ice floes that don't grow as thick as before and which melt away more rapidly come spring.

The insulation value of snow also plays a key role in the maintenance of **permafrost**. About 25% of the Northern Hemisphere land area is underlain by permafrost—ground that remains frozen year-round (Fig. 6-9). The top 50 to 100 cm (20 to 40 inches) of that ground thaws each summer, then refreezes when winter comes back. This is called the **active layer**. As with sea ice, a thick, insulative snowpack

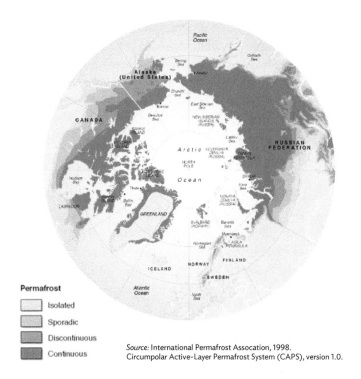

Permafrost

☐ Isolated

☐ Sporadic

■ Discontinuous

■ Continuous

Source: International Permafrost Assocation, 1998.
Circumpolar Active-Layer Permafrost System (CAPS), version 1.0.

FIGURE 6-9. The distribution of continuous, discontinuous, sporadic, and isolated areas of permafrost in the Northern Hemisphere. These areas coincide closely to the tundra and taiga snow climate classes (Figs. 1-4 and 1-5). These are the snow classes where the snow lingers seven to nine months of the year, so snow cover depth, duration, and insulation properties play a large role in determining the state of the permafrost.
(International Permafrost Association. Used with permission.)

tends to retard the winter refreezing process, while little or no snow promotes a deep and rapid freeze-back. Most predictions of future Northern snow conditions for those places where permafrost underlies the landscape suggest deeper snow in mid-winter is coming, and that snow is likely to melt away sooner in spring, exposing the active layer to earlier thawing. Neither of these snow trends is a good thing for maintaining permafrost in the long run.

6.3 SNOWMELT RUNOFF

It is estimated that one-sixth of the world's population (1.2 billion people) relies on snowmelt water for agriculture and human consumption. In the eleven western U.S. states, the mountain snowpack sources between 50% and 90% of all runoff, depending on where in the West you live, and as a result has a value of close to a trillion dollars per year (Fig. 6-10). This snowmelt water is doubly valuable because snow lingers in the mountains and the water runs off in summer when rainfall is scarce and farmers and ranchers need the water most. Even in the largely snow-free areas of California like Los Angeles and San Francisco, people drink snowmelt water and get power from electrical turbines turned by snow water.

Because of its value and the need to apportion the runoff between multiple users with legal water rights, tracking how much snow is stockpiled in basins is done by a number of federal and state agencies in the United States. The U.S. Department of Agriculture/National Resource Conservation Service operates hundreds of manually measured snow courses and over eight hundred automatically recording SNOTEL sites (https://www.wcc.nrcs.usda.gov/snow/)

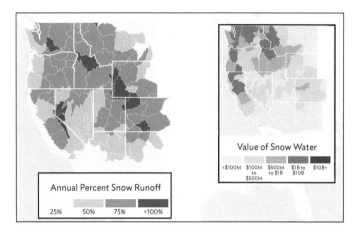

FIGURE 6-10. The annual percentage of runoff from mountain snowpacks by river basins in the eleven western states. (Inset) The dollar ($) valuation of that runoff, using a conservative price of $100/acre foot. Some basins in Utah, Wyoming, and Colorado get 90% of their water from snow.

(*Source:* An interactive map of western U.S. snow basins and the $ value of that water resource can be accessed at https://porkloin.github.io/basinSnow/.)

across the U.S. Additional snow monitoring sites are operated by the U.S. Army Corps of Engineers, the U.S. Geological Survey, and the Bureau of Reclamation. Most of these sites report snow depth, air temperature, and other weather parameters, but most importantly, they report the snow water equivalent (SWE), because this value is a direct measure of how much water that snowpack will produce. Both current and historical data from these sites can be accessed online for free. Not only are these data useful for water supply forecasts, but they are also useful in planning a ski or snowmobile outing where snow conditions matter.

SNOTEL and other automated sites use one of several technologies to measure the amount of snow. The depth is typically measured using a sonic sounder hung from an arm off of a tower. This type of device sends a sonic ping downward and computes the distance to the snow surface based on the time it takes for the echo to return. The depth profiles in Figure 4-3 were recorded using that type of device, which can be purchased from several manufacturers (Campbell, Judd, MaxBotix). Snow water equivalent (SWE) is often measured manually on a monthly schedule using a coring tube (Fig. 6-11) and weighing the core. Various sizes and designs of tubes are available (Federal, Snow-Hydro, ESC). For those locations where SWE is measured automatically and continuously, the device used is essentially an electronic version of a bathroom scale placed under the snow, though in practice these devices have to be considerably bigger, often 3 m (10 feet) on a side. These outdoor scales use a pressure transducer and digital read-out to determine the weight of the snow and are calibrated so that a known pressure can be converted directly into an SWE value. Such devices are still known as *snow pillows*, the name a holdover from the days when the weight of the snow was determined by how much fluid was forced out of a liquid-filled bladder under the snow. That fluid, a mixture of glycol and water, is toxic, and there has been an effort to move to more environmentally friendly gauges. Another less widely used, but a still viable method to measure SWE, is to mount a gamma ray detector above the snow. The difference in gamma count from when the site was snow-free can be converted into SWE.

One widely held misconception is that sites where SWE are monitored for water are "representative" of the basin in which they are located. As Figure 6-11 suggests, the measurement is being made in a nice clearing in a forest, easier

FIGURE 6-11. A field technician is holding an aluminum coring tube called a Federal (or Mt. Rose) sampler. It has a steel cutter head at the lower end. Behind the technician and lying on the snow is a cradle and spring balance (blue). He will core straight down through the snow until the cutter touches the ground, then he will record the depth, then twist the corer into the ground to cut a plug of turf or soil that will hold the snow in the tube. Removing the corer carefully, he will then lay the tube horizontally in the cradle and read the weight from the spring balance, which for this particular device, is calibrated in inches of SWE. (Photo by Matthew Sturm.)

to get to, easier to work in, than the brush and trees back behind the technician. Depending on how many clearings versus tightly packed tree stands there are in the basin, the SWE may or may not be the average for that area. That is not a problem, however, because the value obtained at the site is an index value that is used in models and regression equations to predict the runoff at a stream or river gauge that could be tens of miles away downstream. These models get calibrated using the actual runoff over a period of several years, and once calibrated, do a fine job of predicting the runoff using the index value, just as long as the characteristics of the basin have not changed due to fire, logging, human development, or climate change.

If the amount of snow stored in a basin or watershed is known, then it is possible to predict the runoff as a function of time and the weather (Fig. 6-12). This approach is called **melt modeling**, or alternately, running a **surface energy balance model**. These are typically computer models, and they have been around since the 1970s, used operationally by the Weather Service, the National Resource Conservation Service, and other groups, to predict the daily outflow of meltwater from the snowpack, and to anticipate potential snowmelt flooding.

The basic idea behind these models (suggested by Fig. 6-12) is to input the weather (e.g., the air temperature, cloud conditions, wind speed, etc.), the thermal properties, the albedo, and from these to numerically compute the **excess heat energy** from the sun and air that is available to melt the snow. This excess heat is then turned into an equivalent amount of runoff, which is routed down the mountain and to the reservoir or irrigation ditch. As the snow in a basin melts,

FIGURE 6-12. Modeling runoff starts with computing the snow surface energy balance and converting that energy into the amount of snow melted. (Photo by Matthew Sturm.)

the amount of remaining snow decreases, and the amount of bare ground increases, so the average albedo of the basin starts dropping. This accelerates the pace of the melt, producing runoff more rapidly, but from an ever-decreasing store of snow. The relationship between the air temperature, the melt rate, the amount of snow remaining in the basin, and the runoff are often summarized using a ***depletion curve*** (green curve in Fig. 6-13) coupled to a ***runoff hydrograph*** (blue curve, upper right in Fig. 6-13).

The snowmelt runoff is always delayed getting out of the basin. The peak snowmelt discharge usually occurs after all of the snow in the basin is gone (Fig. 6-13, upper right: the hydrograph). This happens because the meltwater has

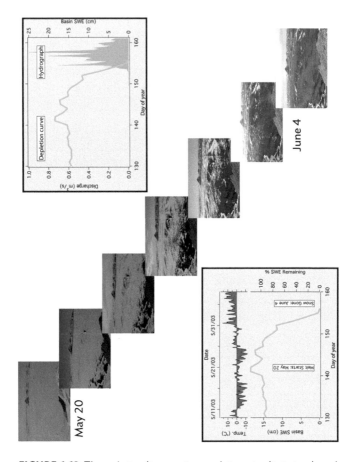

FIGURE 6-13. The spring melt converts snow into water. It starts when air temperatures start going above freezing during the day (red areas in the lower left inset box), and picks up speed when these temperatures remain above freezing both day and night. As the snow melts, more and more bare ground is exposed, reducing the albedo of the basin and promoting an ever-faster melt. The progressive loss of the snow is summarized using a depletion curve (green line), which records how much SWE remains in the basin as a function of time. As the basin nears a snow-free state, meltwater finally starts appearing in the creek or river that comes out of the basin (upper right inset), but there is a significant delay between the start of the snowmelt and when most of the water is flowing downstream.

to percolate down through the snow, recharge the organic layer of the soil under the snow, eventually make its way into a creek or stream, then flow down to a gauging station. Not surprisingly, routing the water down through and under the snowpack remains challenging, with most models having to be calibrated to get this delay right.

6.4 SNOW MECHANICAL PROPERTIES AND TRAVEL

Our lives, and those of the plants and animals that live where there is snow, are affected not only by the thermal and radiative properties of snow, but also by the mechanical nature of the snow—its mass, cohesion, and coefficient of friction—properties that affect how we move over or through the snow. We will see in the next chapter how, through the millennia, plants and animals have adapted to those properties through evolution. Man, the toolmaker, has adapted too, but usually by coming up with clever innovations and tools that take advantage of snow's mechanical properties.

6.4.1 Skis

Thousands of years ago, humans discovered that it was easier to slide something along the top of the snow than to drag it over bare ground. They also realized that if they weren't sinking into deep snow, it was easier to travel. This led to the invention of skis. There are rock drawings of skis dating back more than eight thousand years from the Altai Mountains, and skis preserved in European bogs that may be even older (Fig. 6-14).

By the twelfth century, the ski was already a key part of Nordic culture. In the iconic painting shown in Figure 6-15,

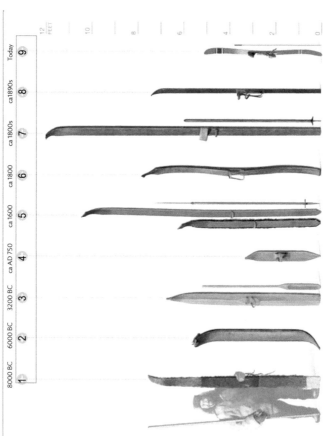

FIGURE 6-14. Skis through the ages have varied in size, form, and the nature of the ski base as humans have tried to maximize the ease with which they could slide over the snow, have sufficient traction to move forward, yet not sink in too far.

(National Geographic, 2004.)

FIGURE 6-15. Two Viking warriors on skis rescue the child-prince Haakon in the year 1206. (www.skiinghistory.org/.../Ski%20Warriors.jpg.)

the two Vikings are carrying two-year-old Prince Haakon to safety in Østerdalen from Lillehammer wearing quite elegant skis, an event supposed to have taken place in 1206. I ski a lot, but have yet to have done so carrying a shield and battle-ax.

While skis have been around a long time, our scientific understanding of how they manage to slide over the snow continues to evolve. That understanding is tied up with a similar problem related to ice skates and ice. In 1886 an engineer, John Joly, put forward the ***pressure melting theory***: that the thin blade of an ice skate puts such enormous pressure on the ice that the it melts due to the pressure-temperature relationship for ice 1h (Chapter 3: the melting temperature of ice is depressed under higher pressures). The pressure-derived meltwater lubricates the blade, and it slides easily over wet ice. Unfortunately, the pressure theory only explains ice skate sliding down to about -4°C (25°F), and cannot not explain ski sliding at all, because the surface area of skis is so much larger than an ice skate blade it cannot put the snow under much pressure at all. Yet ice skating and cross-country skiing both work fine at temperatures down to at least -10°C (14°F).

In 1939, two other scientists, Bowden and Hughes, recognizing the flaw in pressure theory, put forward the ***frictional heating hypothesis***: that the friction of the skate or ski moving rapidly melted little bits of ice, which then lubricated the movement. The chief scientist at the lab where I worked in the 1980s and 1990s, Sam Colbeck, instrumented skis with thermocouples and was able to measure this frictional heating, which indeed takes place. But he also found many other factors at work in producing a lubricating water film, including the material and color of the ski base. He confirmed what most skiers already knew: that there was an optimal water film thickness. Too thin and snow grains poked through the layer and grated on the ski; too thick and water suction dragged on the ski. Newer work using thermal imaging has now called into question whether any frictional melting is taking place beneath a ski (or sled runner). Using this new technology, the scientists found that instead of melting snow grains, ski friction loosened and rolled grains under the ski, or were abraded and broke as the ski traveled over them, both mechanical responses aiding the sliding. Other scientists have questioned this result, so perhaps in the next few years new studies will put this old problem to rest.

As for ice skates, the consensus view now is that they slide because there is a ***liquid-like layer*** present at the surface of all ice, including snow grains, and that the layer

FIGURE 6-16a. Two tiny ice spheres hung on hairs are brought together. They stick due to the liquid-like water film on both balls. When gently tugged apart, they rotate before separating, confirming they have been held together by liquid suction, not freezing.

(U. Nakaya and A. Matsumoto, Simple Experiment Showing the Existence of Liquid Water Film on the Ice Surface, *Journal of Colloid Science* 9, no. 1 [1954], 41–49.)

exists without pressure or friction needed to put it there. The existence of such a film was first suggested by Michael Faraday in 1850, but it took over a hundred years to confirm his suggestion through experiment. In the 1950s the Japanese snow scientist Ukichiro Nakaya, as well as C. L. Hosler in the U.S., proved the existence of this film in an elegant way. They hung tiny ice balls on hair-like threads, then brought the balls together (Figs. 6-16a and 6-16b). The balls stuck, even though the ice temperature was well below freezing. That alone did not clinch the liquid-like layer argument. But next they gently tried to pull the balls apart using the hairs and a tensioning device. The balls first rotated before they finally parted, something not possible at all had they simply frozen together. Today, it is possible using specialized optical microscopy to actually see these molecular-thin films and watch how they evolve,[1] and the fact that there is a water-like film on all ice (including snow grains) is universally accepted. This film, plus undoubtedly other mechanisms, is at the heart of why skis slide well over snow.

[1] (https://www.pnas.org/content/109/4/1052/tab-figures-data)

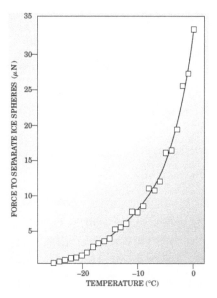

FIGURE 6-16b. The force needed to pull the ice balls apart increases with temperature because the water film thickness increases, as does the cohesive suction.

(From U. Nakaya and A. Matsumoto, "Simple Experiment Showing the Existence of 'Liquid Water' Film on the Ice Surface," Journal of Colloid Science 9 no. 1 (1954): 41–49; and C. L. Hosler, D. C. Jensen, and L. Goldshlak, (1957). "On the Aggregation of Ice Crystals to Form Snow," Journal of Meteorology 14, no. 5 (1954): 415–420.)

6.4.2 Snowshoes

The origin of snowshoes is shrouded in the mists of time, but probably is similar to that of skis. The footwear of Ötzi, the famous iceman found on the Austrian-Italian border, seemed to be designed to engage with a teardrop-shaped snowshoe (Fig. 6-17), which would put the origin of snowshoes back to at least 5,300 years ago. Like skis, snowshoes are designed to increase the surface area over which the weight of a person is distributed and thereby reduce how far they sink into the snow.

FIGURE 6-17. A snowshoe dated to 3,700 BC, about five hundred years before the time of Ötzi the Iceman, suggests Ötzi used snowshoes in his travels.

(https://www.iceman.it/en/snowshoe-found-south-tyrol-italian-alps-older-than-oetzi-the-iceman/.)

Three factors determine sinkage: the weight of the person, the density of the snow, and the degree of bonding or sintering (see Chapter 5 and Fig. 6-18). Sintering (fusing through molecular bonding), is a function of time, temperature, and pressure, and it is likely to be least well developed in a thin, cold snow cover that has seen little wind action, which aptly describes the taiga snow class. There we might expect snowshoe designs that provide the most "float," and indeed historically, that has been the case (Fig. 6-19).

Snowshoe design and craftsmanship reached its apex in North America where indigenous tribes developed designs exquisitely adapted to the local terrain and snow conditions (Fig. 6-20). They also built toboggans, narrow plank sleds designed to be towed behind a person walking on snowshoes

| 35 minutes contact | 279 minutes contact | 1,369 minutes contact |

FIGURE 6-18. Three ice grains sintering at -3°C (27°F). At 1,369 minutes (about a day), they are well sintered and would support a person on snowshoes with limited sinkage (from Kuroiwa, 1961). At a lower temperature, the rate of sintering would be greatly reduced.

(D. Kuroiwa, "A Study of Ice Sintering," *Tellus* 13, no. 2 (1961), 252–59.)

FIGURE 6-19. A set of huge snowshoes designed for the deep, soft snow of the taiga forest.

(Photo by H. Huntington.)

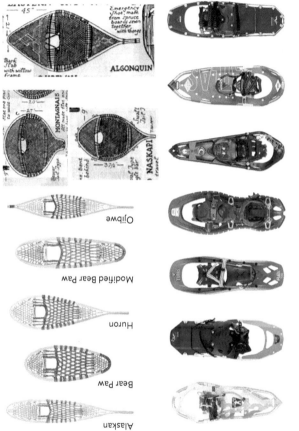

FIGURE 6-20. Various traditional and modern snowshoes. The latter, being smaller but lighter, are often designed for compacted snow trails. (Compiled by Matthew Sturm.)

139

FIGURE 6-21. A traditional towing toboggan. (Photo by Matthew Sturm.)

(Fig. 6-21), a design that maximizes the surface area bearing on the snow and therefore minimizes sinkage. Written references to toboggans date back to the early 1700s, suggesting it also has an ancient origin. The toboggan was primarily designed for use in the taiga, with its soft and deep snow and where those big snowshoes were needed to provide adequate floatation.

Today, dozens of types of sleds are used during oversnow travel. Living in the Arctic and traveling sometimes thousands of kilometers by snowmobile each winter, I have been particularly interested in the sleds people build and pull (Fig. 6-22). There are literally hundreds of variations, with designs and construction ranging from traditional to modern. But even in the most traditional designs today,

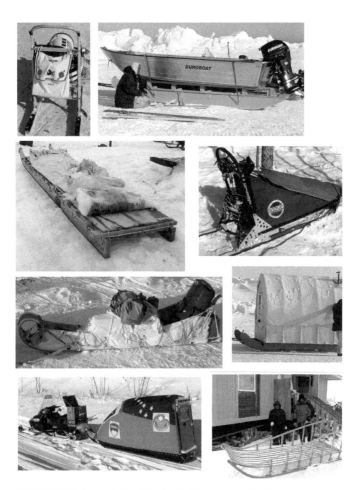

FIGURE 6-22. Some sleds of the Arctic I have seen or built.
(Photos by Matthew Sturm.)

FIGURE 6-23. A Twin Otter aircraft on skis landing at Daring Lake, Northwest Territories, Canada. (Photos by Matthew Sturm.)

UHMW (ultra-high-molecular-weight polyethylene) plastic has pretty much replaced any other material for runners because of its durability and superb sliding ability over snow.

The need to reckon with and adapt to snow mechanical properties has not only led to the development of many types of sleds, but also many other technologies as well, including aircraft on skis (Fig. 6-23) and powerful plows and snow blowers (Fig. 6-24) that can cut through snow drifts in excess of 3 meters (10 feet) deep to clear snow from airstrips, railroads, and roadways. It has also led to a pretty comprehensive understanding of how to keep roads and buildings clear of snow using snow fences that can control where the big drifts are located.

In fact, the most common modern way to handle travel related to snow is to remove the snow and travel as if it were not there, a concept that works reasonably well but costs tens of billions of dollars per year in the U.S. alone, and occasionally

FIGURE 6-24. A powerful snow blower clearing a drifted road.

(like after a blizzard) is not possible. The Snow and Ice Management Association estimates that in 2016, $23 billion was spent by the private sector alone in snow control and removal (both U.S. and Canada), and that these snow removal services employed over 100,000 people. The American Association of State Highway and Transportation Officials (AASHTO) tallied up over a billion dollars in snow removal in just ten states alone for 2014–2015, suggesting costs across all states in the public sector for snow removal of upwards of $5 billion (Table 6-2). Snowshoes and sleds anyone?

So far, I have emphasized how humans have built hardware—from skis and sleds to rotary plows—in order to cope with snow, and I purposely used the phrase "man, the toolmaker" to introduce the section. But of course, women have played as large a role in adapting to snowy environments as men, and one of the more compelling stories in that saga is a story of soft goods rather than hardware. That story starts

TABLE 6-2. Snow removal costs for ten U.S. states, winter 2014–2015

State	Millions $
Pennsylvania	272
Massachusetts	154
Ohio	120
Maryland	108
Michigan	80
New Hampshire	46
Connecticut	45
Kentucky	41
Indiana	40
Colorado	37
TOTAL	$ 943,000,000

Source: Compiled by Matthew Sturm from several sources, including the Snow and Ice Management Association: https://www.sima.org/.

about 27,000 years ago, and it literally balances on the point of a needle. Before humans could migrate from Eurasia to Alaska across the Bering land bridge, they had to learn how to survive brutal Arctic snow conditions and how to avoid freezing to death in fierce Ice Age blizzards. That survival hinged on having weatherproof clothing, and to make such clothing required an eyed needle and advanced sewing technology. We can probably assume somewhere in the mists of the past, it was women who developed eyed needles and then learned how to sew furs exceptionally well.

Eyed needles have been found in Eurasia dating back 35,000 years, and carved bone figurines from southern Siberia suggest that weatherproof fur suits, complete with snug-fitting hoods, were already being produced 24,000

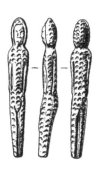 

FIGURE 6-25a. Carved stone figurines from Siberia circa 24,000 years ago showing well-tailored, tight-fitting fur snowsuits with hoods.

(Rasmussen, K. (1927). *Across Arctic America: narrative of the fifth Thule expedition.* GP Putnam's Sons, New York.)

FIGURE 6-25b. An Inuk from Arctic Canada in an exceptionally warm and weatherproof caribou-skin suit.

(Knud Rasmussen's *Across Arctic North America.*)

years ago (Fig. 6-25a). Armed with this snug clothing (Fig. 6-25b), the record suggests humans tried to move into northern Siberia some 20,000 years ago, were pushed back by the bitter conditions of the Last Glacial Maximum, then tried again a few thousand years later. On this second go, they made it.

One of the earliest locations they occupied in Alaska was the Mesa Site, a prominent rock mesa located 34 km (20 miles) north of the Brooks Range. I have stood on this mesa in the dead of winter and wondered how any humans could have successfully lived in such a cold and snowy place. But they did, and they owed their survival not only to hunting and travel hardware like snowshoes and sleds, but also to exceptionally well-sewn clothing.

7

SNOW AND LIVING THINGS

Chapter 6 explored how humans build hardware and soft goods in order to make it easier to live in the snow, but the story of snow and all living things—plants, animals, and humans—is far grander and more complex than just snow tools. Because snow is so widespread on Earth, and has such a strong impact on the natural environment when it is present, snow and life is a massive topic. Here, at best, I can only provide a few examples and hope that these will inspire readers to observe more of this topic on their own.

7.1   ANIMALS

Snow comes in winter, a cold time when plants have senesced (gone dormant) and food is scarce. Animals have evolved three basic strategies to deal with this dark, cold, and snowy period. They can leave for warmer climates (***migration***), stay and **hibernate**, reducing their need for food by lowering their expenditure of energy, or they can **adapt** (through evolution) in ways that allow them to travel through, over, or under the snow and find what food there is for sustenance.

The adaptations include changing color (to white) to be harder to see (and thereby avoid being eaten), having large, snowshoe-like feet in order to "float" on the snow, or conversely, having long legs that make walking in deep snow easier. Reducing heat losses by sporting superb pelage (fur) or downy feathers is another common adaptation to cold and snow. One other clever adaptation is the development of counter-current heat exchange mechanisms in the flow of arterial and venous blood in limbs in order to limit the loss of heat in the extremities. There are also behavioral adaptions (feeding and travel related) that help animals deal with snow and cold.

Some of the animals and birds that turn white include the snowshoe hare, the short-tailed weasel (Fig. 7-1), and the ptarmigan (Fig. 7-2). Other animals like the arctic fox and the polar bear stay white all year round, though in the case of the latter, it is not in order to elude detection by predators. The white of a polar bear, as Figure 7-3 shows, is quite different than the white of snow. It has been my experience when working out on the sea ice that they appear dark against a snow background, making them relatively easy to spot.

Whether white or another color, most winter-adapted animals have fur that is an exceptionally good insulator (Fig. 7-4) and is also beautiful. The ones that lack this good fur have to make up for their lack of insulation by either increasing their metabolism by eating more, or by reducing heat losses by burrowing beneath the snow into a snug den or cavity lined with dry vegetation (shrews, lemmings, weasels). In some cases, they also go into a state of torpor (arctic ground squirrels). Even the arctic fox, with the best pelage of all, uses behavioral adaptations related to snow to help it stay warm. It digs dens in the snow and shelters in these from wind, but if digging is not possible, the fox curls up tightly with its tail across its head, letting the snow drift over it to provide thermal insulation.

The rich fur of these snow- and cold-adapted animals has played a significant role in the history of both the U.S.

FIGURE 7-1. A short-tailed weasel in winter coat.

(http://www.lewis-clark.org/ content/content-article. asp?ArticleID=1906.)

FIGURE 7-2. A willow ptarmigan, the official Alaska state bird, in its winter plumage.

(Photo by Ken Tape. Used with permission.)

FIGURE 7-3. A polar bear stands out against a white snow background.

(https://www.cbc.ca/news/canada/newfoundland-labrador/polar-bear-hopedale-jenna-flowers-1.4632681.)

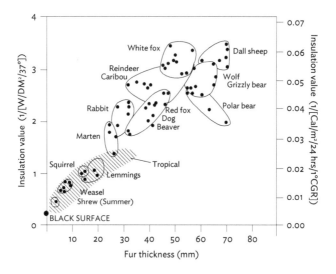

FIGURE 7-4. The insulation value of the pelage (fur) of various arctic animals. The thicker the fur, the better the insulation. The arctic fox has the best insulation, which allows it to survive bitterly cold winter temperatures while eating very little. The animals in the lower left corner, lacking sufficient insulation, live under the snow in winter (gray region on the illustration).

(P. F. Scholander, Vladimir Walters, Raymond Hock, and Laurence Irving, "Body Insulation of Some Arctic and Tropical Mammals and Birds," *The Biological Bulletin* 99, no. 2 (1950): 225–36.).

and Canada, where the fur trade drove settlement and the economy for a period of several hundred years. Fur trading began in Arctic Canada in the 1600s with aboriginal peoples trapping the furbearers and trading the pelts at Hudson's Bay posts for English manufactured goods. From the beginning it was recognized that the colder the climate, the richer the pelt, so the trapping industry pushed west and farther and farther north over time, until it had spanned all of Arctic Canada, all of Alaska, and much of the northern tier U.S. states and Rocky Mountains.

Some animals spend most of the winter living under the snow. For example, in northern Alaska, lemmings (Fig. 7-5a) spend the winter in the subnivean space, covered by a snow quilt that is important to their survival (see also Fig. 6-8). This offers them three distinct advantages. First, they don't need to change color for camouflage. Second, it is warmer and moister than above the snow: while the air temperature above the snow may be -40°C (-40°F), at the base of the snow in a lemming's tunnel it may be only -15°C (5°F), reducing the metabolic cost of staying warm. Third, the tunnels are close to 100% humidity. This is more important than it might seem. In a frozen world, water to drink is precious, and to melt snow to drink takes energy, so living in a moist environment has substantial benefits. It keeps the lemmings from becoming dehydrated, which can happen quickly in the cold Arctic winds. Tunneling under the snow is made easier because of snow metamorphism. The depth hoar at the base of the snow (see Fig. 5-26) is both less dense and more poorly bonded than wind slab or other types of snow layers found higher in the pack, making it an easy layer to dig through.

There are risks for the lemmings (and mice, voles, and shrews, who also live under the snow). While most predators can't fit in subnivean tunnels, weasels, which eat lemmings (Fig. 7-5b), can. It must be terrifying for a lemming to hear a weasel coming. These subnivean prey are also forced out of their snug tunnels on occasion. There have been reports that outgassing of CO_2 from the soil can cause tunnel evacuation, and during snowmelt the tunnels often flood. Then lemmings can be seen running across the snow surface where they are at risk of predation from owls and jaegers (Figs. 7-6a and 7-6b). In spring, jaegers will hover above the snow,

FIGURE 7-5a. A brown lemming.
(www.fr.ch/mhn/labbes/ lemming.htm.)

FIGURE 7-5b. Lemmings, voles, mice, and shrews are at risk when a predator like the weasel gets into their subnivean tunnels. (Matthew Sturm.)

FIGURES 7-6a AND 7-6b. A snowy owl and a long-tailed jaeger.
(Owl: www.dkimages.com/.../previews/835/35001362.JPG. Jaeger: https://birdsna.org/ Species-Account/bna/species/lotjae/introduction.)

peering down, just waiting for an unsuspecting lemming to venture out from its flooded tunnel.

Arctic foxes (Fig. 7-7) also eat lemmings, but have developed a different method of getting at the lemmings under the snow that shows they understand snow layering. With their keen hearing, they can locate the lemmings under the snow. Then they jump in the air and land on the snow, breaking any hard wind slab that might exist and grabbing the lemming from the softer snow below. They know they can break the slab (even though they weigh very little: 3 to 9 kg [7 to 20 lbs.]) because it is underlain by depth hoar that is brittle and will shatter under a sharp blow.

Animals and birds that walk on or through the snow have undergone considerable adaptation of their feet. Specifically, they have developed better fur or feathers to keep these extremities warm and they have developed larger feet so they can walk on top of the snow without sinking in as much. Often the two adaptations have gone hand in hand. For example, the feathery feet of a ptarmigan (Fig. 7-8) stay warmer and provide more float (Table 7-1) than a bird foot lacking these feathery extensions.

FIGURE 7-7. An arctic fox leaps high in order to break through a wind slab and get to prey beneath the slab.
(https://www.youtube.com/watch?v=0EiV-ERKRRs.)

FIGURE 7-8. The foot of a ptarmigan is covered with feathers that keep the foot warm and increase the ability of the ptarmigan to walk on top of the snow.
(Public domain.)

TABLE 7-1. Foot loads for animals that move over the snow

Animal or bird	Foot load (g/cm²)
Snowshoe hare	8
Ptarmigan	15
Lynx	35
Arctic fox	39
Red fox	72
Caribou	400
Wolf	153
Human	180

Source: Compiled by Maria Berger

Perhaps no other animal is more closely associated with snow than the caribou. The vast caribou herds of the North are wonderfully adapted to snowy conditions and thrive as long as the snow is thin enough to allow them to crater through the snow to get at the lichen and plants that they eat. Like the fox, their cratering often relies on their being able to push a wind slab downward into a weaker layer of snow below. Their long legs allow them to move through deep snow rapidly and elegantly and their split hooves spread when loaded by their weight, creating a natural snowshoe (Fig. 7-9) that helps them travel efficiently even in deep snow.

Once, we were measuring snow depth in a remote research location in the mountains when we noticed six caribou grazing on lichen on the top of a small hill. All of a sudden, the caribou charged down the hill toward us, oblivious to our presence. At first, we did not understand why, but then we saw a wolf, a gray female, charging down the hill in pursuit. As Table 7-1 and Figure 7-9 suggest, the caribou have a slight advantage in terms of foot load on the snow over wolves, but a big advantage in terms of leg length. The snow was soft enough that neither the wolf nor the caribou could run on top of the snow: they were all post-holing, but that gave advantage to the caribou with their long legs. The wolf quickly saw that she was not going to catch them and stopped—as did the caribou once they had put about 100 m between themselves and the wolf.

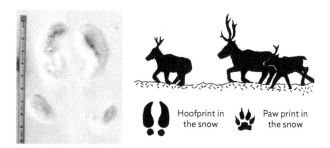

Hoofprint in the snow

Paw print in the snow

FIGURE 7-9. The split hooves and long legs of the caribou make it marvelously adapted to living where there is snow nine months of the year. Its chief enemy is the wolf, whose paws are smaller and sink a little more. (Images by Matthew Sturm.)

Much of the story of snow and animals can be read in snow tracks (Fig. 7-10). Watch for these when you are out:

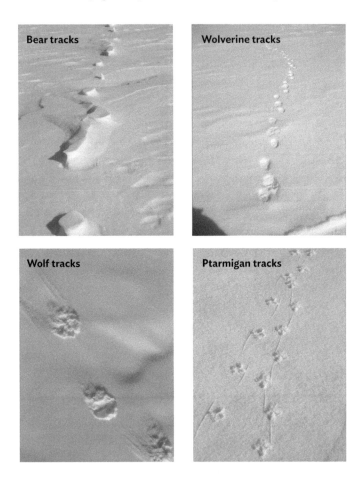

FIGURE 7-10. The bear tracks are raised because the bear packed down the snow, then the wind eroded the snow where it wasn't packed.

(Photos by Glen Liston. Used with permission.)

7.2 SNOW AND PLANTS

The interaction of snow and plants depends on how tall and how supple the plants are, the prevailing snow depth, and whether there is much wind or not. In Chapter 1 the climate classes of snow were introduced, four of which bear the names of vegetation zones (tundra, boreal forest, prairie, and montane forest). This is not just a coincidence (Fig. 7-11). Both the snow characteristics and the vegetation arise from the same climatic forces: the amount of precipitation, the temperature, the progression of the seasons, and how harsh the location is (with wind a good measure of harshness). Of course, the relationship between snow class, vegetation, and climate class cuts both ways. For example, where there are no trees, there is more wind, but where there is more wind,

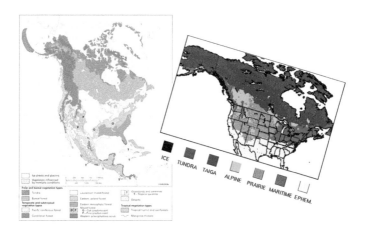

FIGURE 7-11. (Left) Vegetation zones of North America versus (right) climate classes of snow in North America. I have rotated the snow class map to help with a visual match: the similarity in the pattern of the zones and the classes is striking.

snow conditions are harsher so trees may not grow. Without trees, but with wind, a tundra or prairie snow cover develops, rather than a taiga snow cover.

7.2.1 Tundra and Alpine Plants

For climate classes where the snow can drift, Figure 7-12 suggests how tundra montane forest plants and snow vary locally with topographic position. In drift areas the plants are generally buried by snow and therefore protected from the cold, from desiccation, and from abrasion by saltating snow grains. The protection is helpful, but the plants have to pay a price for this luxury. They get a shorter growing season due to the longer time it takes the drift snow to melt out in spring.

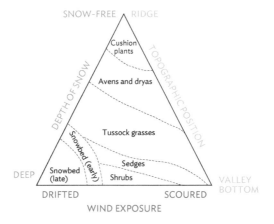

FIGURE 7-12. The relationship between snow, wind, and topographic position for tundra and alpine plants.

(Redrawn from William Dwight Billings and Harold A. Mooney, "The Ecology of Arctic and Alpine Plants." *Biological Reviews* 43, no. 4 (1968): 481–529.)

The other extreme, windswept ridges, are places where only the hardiest cushion plants can survive because there is no snow to protect the plants in winter, though in these locations plants get ample sunlight.

7.2.2 Snow and Shrubs

The interaction between snow and shrubs (willow, dwarf birch, and alder) is particularly complicated (Fig. 7-13) because the shrubs can (a) be bent down under a snow load; (b) trap blowing snow like a snow fence; or (c) if large and stiff enough, intercept snow in the same way as a tree canopy. When the shrubs function as a snow fence, they trap wind-blown snow and create drifts around and immediately down-wind of where they are growing (Fig. 7-14). This deeper snow,

FIGURE 7-13. These willow shrubs are intercepting snow, and, due to the weight of the snow, are bending and being pushed down and buried. They are quite supple, and observations suggest the bending does little harm. With enough snow loading, the shrubs could become completely buried, but if there is a brief thaw or some wind, then the snow-load may come off and the shrubs will rebound, ready to intercept snow in the next snowstorm. (Photo by Matthew Sturm.)

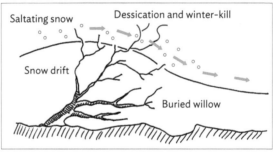

FIGURE 7-14. These willows have trapped saltating snow that was being blown by the wind, creating a drift around and on the downwind side of where the branches are most dense. The ground area covered by the snow is likely to be much warmer than away from the shrubs in the nondrifted snow, particularly on the upwind side of the shrubs where the snow has been scoured away. Any exposed branches on these shrubs are at risk of being abraded, desiccated, and winter killed. (Images by Matthew Sturm.)

being a good thermal insulator, keeps the ground warmer in these areas, and it is thought that this may promote the production of more nutrients, as the soil microbes stay active longer into the winter in the warmer soil. More nutrients means the adjacent shrub grows better next summer, so a positive feedback system is set in operation between snow and shrubs.

The shrubs also contribute to earlier snowmelt because they poke up through the snow and reduce the local albedo (Fig. 7-15). The shrub branches are dark and tend to absorb solar radiation. Small cavities and chimneys form in the snow around the branches when the sunlight warms them, a process that seems local and unimportant until we realize that there are billions of these little shrub-induced melt accelerators operating during the melt.

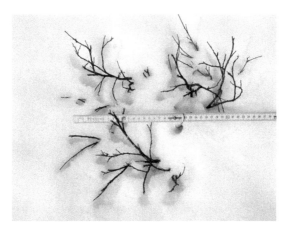

FIGURE 7-15. Dark shrub branches sticking up through the snow absorb solar energy and accelerate snowmelt wherever the shrubs protrude from the snow. (Photo by Matthew Sturm.)

7.2.3 Snow and Trees

The quintessential Christmas card seems to be of a spruce or fir tree heavily loaded with snow (Fig. 7-16), and this association (snow and trees) is not misplaced. The two have an intimate relationship with one another, the trees impacting where the snow lodges and lies, the snow determining what moisture the trees will have to grow.

Much of our knowledge of the interaction of snow and trees, particular conifers, has its early beginnings in a bitter fight over water rights in California. This was not the more well-known fight over water for Los Angeles (e.g., the movie *Chinatown*); it took place farther north near Lake Tahoe. In 1903 a scheme was put into operation to dam the Truckee River, increasing the height of the lake and therefore ensuring a greater summer outflow. This water would then be used for a big agricultural project over in Nevada. The scheme

FIGURE 7-16. A classic Christmas card with spruce trees laden with snow. Such images conjure up tranquility, beauty, and peace, but the process of loading the branches is anything but simple.

made few people happy. The dam raised the water level of Lake Tahoe, flooding the waterfront properties of many rich and well-connected individuals, while the radical year-to-year fluctuations in the outflow meant that the water supply to the farmers was unreliable. Then came the winter of 1906–1907 with a record 22 m (72 feet) of snowfall in the Sierra Nevada. The water behind the dam reached dangerous levels so the operators were forced to spill water in spring, before it could be used by the farmers. Predictably, by summer, drought conditions prevailed, but now there was no more water behind the dam for the crops.

Enter Dr. James Church, the inventor of the Mt. Rose (or Federal) snow sampler (Fig. 6-11). Church realized that the water being battled over was all from snowmelt, and that unless the amount of snow could be measured prior to the melt, no rational operation of the dam, the power plant, the irrigation project, or lake level could take place.

So what does this battle over water rights have to do with trees? One other group was quite active at this time: the loggers, and Church realized that it mattered whether the Sierra snow fell on trees or into large clearings. Snow that fell on tree branches was subject to much higher rates of sublimation losses than snow on the ground. The loggers were rapidly turning the slopes surrounding Lake Tahoe from forests to clearings. But how much more snow remained when there were no trees . . . and how big could a clearing (or logging area) be before the wind got into it and drifted and sublimated the snow away?

Trying to solve that problem—the problem of interception of snow by tree branches, and the fate of that snow depending on where it lodges—had its scientific start with

Church's historic studies, but it continues today because the answers are complex and differ depending on the nature of the trees, the type of the snow class that blankets the forest, and the impact of fire and insects on that forest as well. Church's main contribution was drawing attention to the fact that in order to answer such questions objectively, we needed to institute methodical measurement programs. These had their start then and have been institutionalized in the network of snow courses and SNOTEL sites operated through the Western U.S. today by the USDA National Resource Conservation Service.[1]

Snow falling on and through the branches of a tree is a filtration process, though in time the "filter" becomes clogged and the way in which it filters the snow changes (Fig. 7-17). At first, when the branches are bare, as snowflakes hit them, the flakes bounce easily (because snow bouncing on wood or needles is a fairly elastic process), and many snowflakes make it through the branches to the ground. At this stage the process of interception is slow. Eventually, some snowflakes begin to lodge in the needles, and particularly if these snowflakes are stellar dendrites, they will provide a better, stickier target for the next snowflakes that fall. This next wave of snowflakes can now cohere to, or interlock, with the snowflakes already in place (Fig. 7-18). The interception efficiency rises quickly. At some point, there will be enough snowflakes to bridge most of the "holes" in the canopy (areas where looking straight up one would see light), and the trapping efficiency will reach a maximum. Additional snow will start to weigh down the branches, which will bend, and now these steeper snow surfaces will once again pro-

[1] (https://www.wcc.nrcs.usda.gov/snow/snotel-wereports.html.)

mote the bouncing of snowflakes off the branches and out of the canopy (Fig. 7-17). If the branches dip enough, they may even unload. The sloped branches, even if still snow covered, are no longer as optimal an interception target as before, and the rate of interception now drops. The snow will stay in the branches until either a wind comes up and shakes the snow loose, or the temperature rises to above freezing and liquid water lubricates the intercepted snow, allowing it to slide off. Then the process begins again in the next snowstorm.

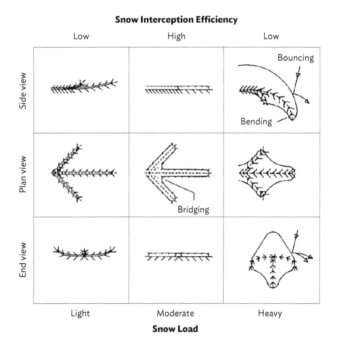

Snow Interception Efficiency

FIGURE 7-17. Three stages of snow interception efficiency (low—high—low) by conifer tree branches.

(After R. A. Schmidt and David R. Gluns, "Snowfall Interception on Branches of Three Conifer Species." *Canadian Journal of Forest Research* 21, no. 8 [1991]: 1262–69.)

FIGURE 7-18. Spruce needles choked with newly fallen stellar dendrites, which are now interlocking with other dendrites and beginning to bridge from one branch to another. (Photo by Matthew Sturm.)

A student of mine spent three winters monitoring the loading and unloading of spruce trees by snow using a terrestrial lidar, a survey device that allowed him to create a 3D movie of trees over time (Fig. 7-19). When we looked at the results, it was as if the trees were breathing, the branches rising and falling as the snow loaded them, was shed, then loaded them again. It was a strange perspective: the trees seeming more mobile and alive than we typically think of for plants. But trees come in so many forms, shapes, and sizes— deciduous tree, conifers, tall trees, small trees, bent trees— that the interaction of snow and trees takes on as many personas as the trees themselves, as suggested in Figure 7-20. I think it may be many years before snow scientists have a universal model of snow and tree interactions.

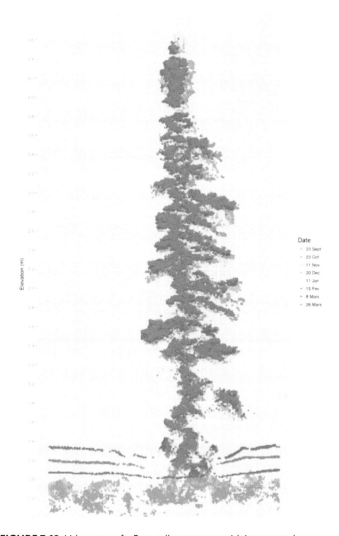

FIGURE 7-19. Lidar scans of a 5-m-tall spruce tree with intercepted snow load from eight times during the winter. The different colors capture how the branches collectively moved up and down as the loading changed. (Courtesy of S. Filhol.)

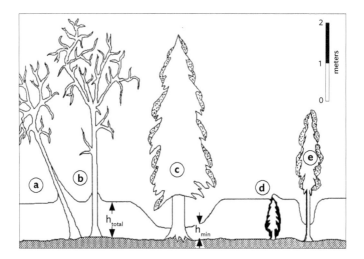

FIGURE 7-20. The figure suggests the wide range of ways snow and trees can interact. The deciduous trees (a and b) have cones of snow about them because against the trunk the snow does not settle well, while farther away it does. The depression under the large conifer (c) is called a tree well, and it is there because the snow that was meant to be there has lodged in the tree canopy. In deep snow packs, these wells can be dangerous—hard to get out of if one accidently falls in. The ratio of the tree height to the snow depth matters as well. If the snow is deep enough, then it can inundate small trees (d and e) and lay down most shrubs.

7.3   SNOW AND PEOPLE

I suspect that every person who has ever experienced snow, even briefly, develops his or her own unique relationship with it. I was hiking in Scotland on the Isle of Skye, and it had snowed a few centimeters. A Brazilian mother and her son were hiking nearby. The boy had never seen snow before, yet he immediately flopped down in the snow and made a snow angel, a smile a mile wide on his face. Why? And how did he even know about this strange human ritual? Bernard Mergan in his book *Snow in America* uses historical sources to show that snow has played a key role in the development of the American psyche. He makes the case that snow has always elicited strong emotions, both positive and negative, throughout U.S. history, and that dealing with winter and harsh snowy conditions has made Americans tough. Perhaps. It is easy to agree with him that snow affects us all in deep ways, though I am certain these experiences are not limited to the U.S.

That snow lies deep in our psyche is suggested by several somewhat contradictory points with which most people would agree:

- Snow is often used to symbolize cleansing.
- Snow, with its blindingly bright white color, is often used to denote purity and innocence.
- Yet at the same time, snow and winter are also often used to represent sadness, bleakness, or death.
- Yet again, winter is often depicted as a cozy family time (think Christmas and Grandma Moses's paintings).
- And yet still again, snow and winter are often depicted as evil (or at least frigid): think *The Snow Queen* and *Frozen*.

These points seem to suggest that while we are conflicted about snow, it indeed plays a powerful role in both our physical and mental lives.

There are more quantitative measures that suggest the nature of our relationship to snow. In 2013 there were 144,600 snowmobiles sold worldwide. At $6,000/machine, that is more than a billion dollars in sales. If we include the associated costs of snowmobile recreation (special clothing, lodging, trailers to move the machines, groomed trails, etc.), it is estimated that this winter recreation alone is over a $20 billion/year endeavor worldwide. Moving on to skiing, it is estimated that there are 330 million skiers worldwide, and that skiing (mostly downhill skiing) is a $14 billion a year industry. Vail Ski Area alone averages 1.6 million skier visits per year. Add to this the billions spent to clear snow (see Chapter 6, Table 6-2), and the hundreds of billions of dollars that snow water is worth, and it is clear snow is not only an essential part of our lives, but also an essential part of the global economy.

So what is the future of snow? What can we expect? A writer friend of mine some years ago approached me. She wanted to write a book about the ***demise of snow***. She lived in a place where rain-on-snow was already becoming more common, and the trend toward more rain and less snow (due to warmer winter temperatures) looked like it might end up meaning no-snow winters for her. What did I think? My answer was to invite her to come on one of our Arctic snow expeditions. Snow in the Arctic may be changing, but it is not going to disappear for a very long time. That helped her, I think, because she was upset at the thought of losing snow, though it didn't answer her specific question.

Currently, my synthesis of the outlook for snow globally is this: there are four negative snow trends that have been underway for a few decades, and these are likely to continue. I suspect they will have both positive and negative societal impacts, though perhaps due to human psychology, the negatives may seem to dominate:

Global snow-covered area, which has been declining slightly for the past thirty years will continue to decline, further reducing the ability of snow to cool the planet. The albedo feedback effect from this reduction could accelerate the loss even more and play a significant role in future global warming.

Total worldwide snow volume (or mass) is also thought to have been declining, though our data on this is rudimentary. This reduction shows up as a loss of glaciers and mountain snow, which of course, reduces available runoff in summer. That loss could prove a serious blow to agriculture in the western U.S. states. It is a loss that has been heavily reported in the news. Similar snow volume losses on Arctic sea ice appear to be hastening the loss of that ice, also a topic that has been in the press frequently. Causal analysis suggests these losses are a direct consequence of global warming and likely to continue.

Snow that falls is melting sooner, producing earlier stream runoff over a shorter period of time. Early snowmelt further reduces the time-integrated cooling power of the snow albedo effect and contributes to the feedback power of that process. The earlier runoff is also problematic because with limited reservoir storage capacity, much of the precipitation that used to come as snow, but now comes as rain, cannot be stored and then used later for agriculture. This shortening of the winter season also threatens the bottom line for some ski areas.

In many places, particularly those with more maritime climates, winter precipitation is arriving increasingly as rain, a trend associated with increasing flooding risk in mountain areas. The snow-to-rain transition is having an adverse impact on wildlife (foraging) and transportation (icy roads) as well producing more dangerous travel conditions, and, like the shortening winter season, has forced the closure of a number of lower-elevation ski areas. This snow-to-rain transition is also adversely affecting snow-related water supplies.

7.4   FINAL THOUGHTS

Yet none of this should suggest to the reader that snow will disappear . . . it will just be found in different locations at different times. The snow climate classes might undergo a latitudinal shift, but they, too, will continue to exist. Such changes have occurred in the past. Perhaps as observers, we will see more melt features in the snowpacks of the world, but I am confident that the opportunities to observe and engage with snow will still be widespread. And while it will not solve our climate change–induced societal problems, the magic of this amazing substance, snow, will still be there for anyone willing to look.

FURTHER READING

There are hundreds of books on snow and thousands of technical papers can be found by putting "snow" into a Google search. On avalanches alone dozens of books and guides are available. Below are listed a few sources of material for the reader who wishes to learn more.

Avalanches

Snow Sense: A Guide to Evaluating Snow Avalanche Hazard (paperback), by J. Fredson, D. Fesler, K. Birkeland (eds.). 2011.

The Avalanche Handbook, by Dave McClung and Peter Schaerer, third edition. 2006.

Snow, Weather, and Avalanche Guidelines (SWAG), U.S. American Avalanche Association.

Snowflakes and Snow Crystals

Ken Libbrecht's Field Guide to Snowflakes, by K. G. Libbrecht. Voyageur Press, 2016.

"Snow Crystals," by F. C. Frank. *Contemporary Physics* 23, no. 1 (1982): 3–22.

The Snowflake: Winter's Secret Beauty, by K. Libbrecht and P. Rasmussen, 1134–1135. 2004.

Snowflakes in Photographs, by Wilson Bentley. Courier Corporation, 2012.

Snowflakes: Natural and Artificial, by Ukichiro Nakaya. Harvard University Press, 1954.

Snow and Ecology

"The Ecology of Arctic and Alpine Plants," by William Dwight Billings and Harold A. Mooney. *Biological Reviews* 43, no. 4 (1968): 481–529.

Life in the Cold: An Introduction to Winter Ecology, by P. Marchand. UPNE, 2014.

Snow Cover as an Integral Factor of the Environment and Its Importance in the Ecology of Mammals and Birds, by A. N. Formozov. 1946, original Russian. English translation by the Boreal Institute, University of Alberta, Edmonton.

Snow Classification

"Fall Speeds and Masses of Solid Precipitation Particles," by J. D. Locatelli, and P. V. Hobbs, *Journal of Geophysical Research* 79, no. 15 (1974): 2185–2197.

Field Guide to Snow Crystals, by Edward R. LaChapelle. 1969.

"A Global Classification of Snow Crystals, Ice Crystals, and Solid Precipitation Based on Observations from Middle Latitudes to Polar Regions," by K. Kikuchi, T. Kameda, K. Higuchi, and A. Yamashita. *Atmospheric Research* 132 (2013): 460–472.

International Classification for Snow on the Ground, by S. Colbeck (chair), E. Akitaya, R. Armstrong, H. Gubler, J. Lafeuille, K. Lied, D. McClung and E. Morris. International Commission on Snow and Ice of the International Association of Scientific Hydrology, 1990, http://home.chpc.utah.edu/~u0035056/oldclass/3000/Seasonal_Snow.pdf.

International Classification for Seasonal Snow on the Ground, volume 5, by C. R. L. A Fierz, Richard L. Armstrong, Yves Durand, Pierre Etchevers, Ethan Greene, David M. McClung, Kouichi Nishimura, Pramod K. Satyawali, and Sergey A. Sokratov. UNESCO/IHP, 2009.

"Meteorological Classification of Natural Snow Crystals," by C. Magono and C. W. Lee. *Journal of the Faculty of Science,* Hokkaido University 2, no. 4 (1966).

Physical Studies on Deposited Snow, by Zyungo Yoshida. Hokkaido University Institute of Low Temperature Science, 1955.

"Radiative Forcing and Albedo Feedback from the Northern Hemisphere Cryosphere between 1979 and 2008," by M. G. Flanner, K. M. Shell, M. Barlage, D. K. Perovich, and M. A. Tschudi. *Nature Geoscience* 4, no. 3 (2011): 151–155.

"Studies of snow transport in low-level drifting snow," by Daiji Kobayashi. *Contributions from the Institute of Low Temperature Science* 24 (1972): 1–58.

Snow and Water

Cadillac Desert: The American West and Its Disappearing Water, by Marc Reisner. Penguin, 1993.

The History of Snow Survey and Water Supply Forecasting: Interviews with U.S. Department of Agriculture Pioneers, No. 8, by D. Helms, S. E. Phillips, and P. F. Reich. U.S. Department of Agriculture, Natural Resources Conservation Service, Resource Economics and Social Sciences Division, 2008.

"Snowfall Interception on Branches of Three Conifer Species," by R. A. Schmidt and David R. Gluns. *Canadian Journal of Forest Research* 21, no. 8 (1991): 1262–1269.

Snow Science

Handbook of Snow: Principles, Processes, Management, and Use, by Donald Maurice Gray and David Harold Male (eds.). Pergamon, reissued, 1981.

Snow and Society

The Brothers Vonnegut: Science and Fiction in the House of Magic, by Ginger Strand. Farrar, Straus and Giroux, 2015.

Snow, by Ruth Kirk. William Morrow and Company, 1978.

Snow in America, by Bernard Mergen. Washington, D.C.: Smithsonian Institution Press, 1997. xxi + 321 pp.

Note: page numbers followed by *f* and *t* refer to figures and tables respectively.